CHEMISTRY AND PHYSICS OF CARBON

Volume 22

CHEMISTRY AND PHYSICS OF CARBON

A SERIES OF ADVANCES

Edited by

Peter A. Thrower

DEPARTMENT OF MATERIALS SCIENCE AND ENGINEERING
THE PENNSYLVANIA STATE UNIVERSITY
UNIVERSITY PARK, PENNSYLVANIA

Volume 22

MARCEL DEKKER, INC.   New York and Basel

The Library of Congress Cataloged the
First Issue of This Title as Follows:

Chemistry and physics of carbon, v. 1-
    London, E. Arnold; New York, M. Dekker, 1965-

    v. illus. 24 cm

    Editor: v. 1- P. L. Walker

    1. Carbon. I. Walker, Philip L., ed.

QD181.C1C44   546.681

Library of Congress   1   66-58302
ISBN 0-8247-8113-9

MARCEL DEKKER, INC.
270 Madison Avenue, New York, New York 10016

Current printing (last digit):
10 9 8 7 6 5 4 3 2 1

Preface

This volume of *Chemistry and Physics of Carbon* begins with a chapter that many members of the carbon community worldwide have been awaiting with much anticipation. Dr. Agnes Oberlin has long been recognized for her pioneering electron microscope studies of a variety of carbonaceous materials, and in this chapter she provides us with an up-to-date account of her studies as they relate to carbonization and graphitization. I am sure that this contribution will generate great interest.

In Chapter 2, Drs. Jüntgen and Kühl turn our attention to a completely different subject—carbon as a catalyst. Whereas the title of the paper indicates a very specific area of application, viz. flue gas cleaning, the general principles and science involved will be of interest to many carbon scientists whose work seems far removed from the real world of flue gases.

The final chapter returns us to the subject of adsorption on carbons, a subject that was considered extensively in the last volume of this series and is the focus of much current research on carbon materials. Drs. Jaroniec and Choma have written a chapter to complement that published in Volume 21, but I am sure there is still much to be said on the subject.

I write this as we are preparing for the 19th Biennial Conference on Carbon to be held here at The Pennsylvania State University. So far we have received a record number of contributions. The worldwide interest in carbon is undiminished and appears to once again be on the increase. Our current contributors, from France, West Germany,

and Poland bear testimony to this fact. I trust that the volume will
also be of interest to readers worldwide.

Peter A. Thrower

Contributors to Volume 22

Jerzy Choma   Institute of Chemistry, Military Technical Academy, Warsaw, Poland

Mieczyslaw Jaroniec   Department of Theoretical Chemistry, Institute of Chemistry, M. Curie-Sklodowska University, Lublin, Poland

Harald Jüntgen   Department of Coal Chemistry, Bergbau-Forschung GmbH[*] and University of Essen, Essen, Federal Republic of Germany

Helmut Kühl   Department of Coal Chemistry, Bergbau-Forschung GmbH, Essen, Federal Republic of Germany

Agnes Oberlin   Laboratoire Marcel Mathieu, CNRS-Université, Pau, France

---

[*]Retired.

Contents of Volume 22

3   THEORY OF GAS ADSORPTION ON STRUCTURALLY HETEROGENOUS
    SOLIDS AND ITS APPLICATION FOR CHARACTERIZING
    ACTIVATED CARBONS   197

Mieczyslaw Jaroniec and Jerzy Choma

Contents of Other Volumes

CHEMISTRY AND PHYSICS
OF CARBON

Volume 22

1

High-Resolution TEM Studies of Carbonization and Graphitization

AGNES OBERLIN

Laboratoire Marcel Mathieu, CNRS-Université, Pau, France

|---|---|---|---|
I.	Introduction		2
II.	Image Formation		6
	A.	Lattice Fringes	6
	B.	Dark-Field Modes	8
	C.	Conclusion	31
III.	Spherical Configurations		32
	A.	Concentric Textures	32
	B.	Radial Spherical Texture	55
	C.	Statistically Spherical Textures	65
IV.	Cylindrical Configurations		115
	A.	Filamentous Carbons	115
	B.	Carbon Fibers	122
V.	Conclusion		127
VI.	Appendix		128
	A.	Introduction	128
	B.	Reciprocal Space of Turbostratic Layer Stacks Oriented in Parallel	129
	C.	Reciprocal Space of a Fiber	131
	D.	Reciprocal Space of a Complete Random Sample	132
	E.	Graphitization	133
	F.	Conclusion	135
	References		135

1. INTRODUCTION

When considering the tremendous amount of TEM studies carried out on
carbons for about 20 years, a leading thread seems at first glance
difficult to find. However, mechanical, rheological, electronic, and
structural properties of carbons, among others, depend mainly on the
three-dimensional arrangement of the aromatic layers forming the carbon skeleton. Since the physicochemical properties so strongly depend
on the texture or microtexture, "blind" statistical techniques such as
x-ray diffraction or Raman spectroscopy have to be complemented by
direct imaging, either at the micrometric level (optical microscopy)
or on the atomic scale (transmission electron microscopies). All
carbonaceous materials are poorly organized, so that the transmission
electron microscope or microscopy (TEM) and scanning electron microscope or microscopy (SEM) are powerful tools as imaging and even analytical techniques (owing to x-ray microanalysis, i.e., energy-
dispersive x-ray spectrometry, EDS).

In fact, all carbonaceous materials are initially made of similar
elemental bricks arranged differently relative to each other. The
elemental unit or basic structural unit (BSU) is made of planar aromatic structures consisting of less than 10-20 rings and piled up
more or less in parallel by two to four. It will be shown in this
chapter that the various possible configurations for associating BSU
in space follow only two different symmetries: either spherical or
cylindrical. All possible microtextures derive from these two basic
arrangements by considering variable radii of curvature r of the
sphere or cylinder more or less statistically distributed.

Figure 1 [1-3] sketches the possible microtextures. Figure
1(a,b,c) shows a spherical symmetry, either true or statistical. The
occurrence of an infinite radius of curvature gives rise to flat
lamellae [Fig. 1(d) and (e)]. Figure 1(f) and (g) represent the
cylindrical symmetry, either true or statistical.

Inside each configuration, the aromatic layers can be highly
distorted. Very small BSU are then associated edge to edge with tilt
and twist boundaries. This is the case for most low-temperature

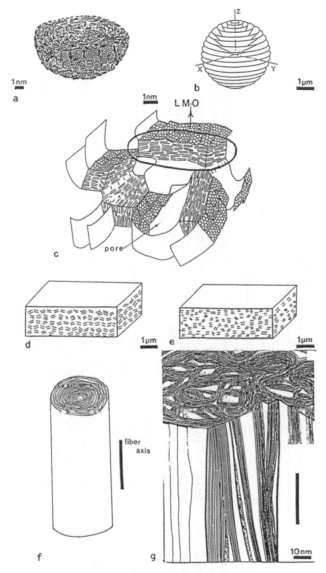

FIG. 1 Sketches of the possible arrangements of BSU. (a and b) True spherical symmetry, with (a) concentric texture (from Ref. 1) and (b) radial texture (from Ref. 2). (c, d, and e) Statistical spherical symmetry, with (c) crumpled sheets of carbon (local molecular orientation or LMO) (from Ref. 3), (d) lamellar texture (infinite radius of curvature or infinite LMO), and (e) lamellar texture with a long-range statistical orientation. (f and g) Cylindrical symmetry, with (f) true cylindrical symmetry and (g) statistical cylindrical symmetry.

TABLE 1 Possible Symmetries in the Three-Dimensional Arrangement of BSU.

Spherical symmetry

True spherical		Statistically spherical	
Concentric	Radial	Soft carbons, thin carbon films	Rough laminar pyrocarbons
Carbon blacks	High-pressure carbonized softening materials	Heavy petroleum products	Smooth laminar pyrocarbons
Carbon shells		Oxidized pitches	Isotropic pyrocarbons
		Coals, kerogens	High-pressure graphitized carbons
		Saccharose, cellulose-based carbons	Anthracites, semigraphites
		Glassy carbons	

Cylindrical symmetry

True cylindrical	Statistically cylindrical
Hollow fibers	PAN-based carbon fibers
Filamentous carbons	Pitch-based carbon fibers

materials. All intermediates are known between the extrema. Some
examples of the carbons known to obey these models are given in
Table 1.

The anisotropy of a pair of aromatic layers is tremendously
large, since the carbon-carbon bonding is the shortest known (0.142
nm), even shorter than the diamond one (0.154 nm), whereas the interlayer spacing is always larger than the graphite d_{002} spacing (0.335_4
nm). The intrinsic properties of a single BSU are thus highly anisotropic. However, the manner in which BSU are associated can enhance
or, on the contrary, reduce or modify this anisotropy. Bulk isometric
properties are to be expected if the symmetry is spherical, except for
an infinite radius of curvature (lamellar texture) where planar anisotropy reaches its maximum. On the other hand, cylindrical symmetries
(fibers) lead to very special tensile mechanical properties along the
fiber axis.

The main interest of imaging techniques such as electron microscopies is to determine and if possible to quantify the relation
between microtexture and physicochemical properties, to make it
possible to predict the latter.

Section II of this chapter discusses from a practical standpoint
how images given by various modes of TEM are formed (disregarding the
simple bright-field images) and which data are obtained from them.
Then in Sections III and IV, respectively, spherical and cylindrical
configurations will be described. Eventually, TEM data will be compared to those obtained from other techniques: elemental analysis
(EA), optical microscopy (OM), infrared (IR) and Raman (R) spectro-
scopies, and x-ray diffraction (x-ray). Finally, the correlation
between microtexture (sometimes also improperly called microstruc-
ture) and material properties will be briefly reviewed. [*Note:* For
correctly understanding TEM imaging, the reciprocal space of all
possible microtextures should be known precisely. A reminder of
these crystallographic data is given in Section VI.]

II.   IMAGE FORMATION

A.   Lattice Fringes

1. *Principle*

In Vol. 14 of this collection [4], the theory and history of the resolution of crystal lattice planes, i.e., lattice fringe imaging (LF), is excellently discussed by Millward and Jefferson. Hence it does not seem necessary to detail this subject further. However, a few additional remarks must be added here concerning numerical data that can be obtained. The reliability of the images and the reliability of the data will be discussed.

2. *Numerical Data*

From the 002 LF micrographs [see Fig. 12(c)], the length of a perfect fringe (L_1) and in some instances the length of a distorted fringe (L_2) can be measured. The number N of fringes in a stack can also be measured. Using a laser beam optical bench, optical diffraction patterns (ODP) can be obtained [1,3]. From these, the relative misorientation of fringes (twist β between layer stacks) and the interfringe spacing D_{002} as well as its fluctuation ΔD_{002} can be measured. In any case, the precision of these measurements entirely depends on the accuracy with which magnification is determined and thus depends on the internal standard used. As discussed elsewhere [4], it is necessary first to know the value of focusing relative to the transferfunction plateau. Except in a thin section, the real level of the object relative to the amorphous area used for determining the transfer function is not known with certainty. Another uncertainty is introduced by the internal standard, which may not be located at the proper level, so that magnification is not known to better than a few percent.

3. *Reliability of the Images*

As shown elsewhere [4], the relation between fringes and the corresponding family of lattice planes depends on the transfer function of the lens. For materials such as carbons, the resolution of the TEM practically limits the efficiency of LF to the imaging of 002 fringes (aromatic layers seen edge-on). From graphite down to low-temperature carbonaceous materials, the mean interlayer spacing $\overline{d}_{002}$ varies from 0.335_4 nm up to more than 0.4 nm, so that reliable imaging is possible only if the plateau of the transfer function includes all possible d_{002}. The authors of Ref. 4 thus demonstrate clearly (compare Figs. 21 and 23 in Ref. 4) that the EM 300 Philips microscope is unable to image with the same underfocus value both graphite fringes and a two-layer-thick particle. For carbon studies, a microscope having a larger plateau in the vicinity of the Scherzer optimum has to be used (EM 400 or CM 12 Philips, for example).

4. *Reliability of the Data*

Practically, only 002 fringes are easily obtained with most of the TEMs commercially available. Many authors were tempted to establish models of carbonaceous materials entirely based on such images. However, it is of utmost importance to keep in mind that "things are seldom what they seem." The great majority of such models are in fact entirely unreliable. A simple reasoning shows that only the part of a carbon layer stack that fulfills at least approximately the 002 Bragg condition is imaged (θ_{002} of graphite = 5×10^{-3} rad, that is, 0.29° at 120 kV). If the layers are distorted, even weakly, the 002 scattered beam promptly disappears and therefore so does the image. Though its image disappears, the stack may be largely extended obliquely relative to the observation plane (Fig. 2). It will be seen further that this feature can be one of the explanations of carbon black LF images [5].

To distinguish unambiguously between a large-diameter isometric carbon layer stack and a ribbon it is necessary to keep in mind that TEM images are two-dimensional projections on the observation plane along the objective lens optical axis. As a result, many details remain unseen unless the object is tilted so that other regions of the distorted layer stack progressively fulfill the Bragg condition. Nevertheless, the resolution of a goniometer stage microscope is generally not sufficient for performing such an experiment. Fortunately, there is another way to obtain other bidimensional projections of the object along different crystallographic directions: wide use of dark-field (DF) images [3,6].

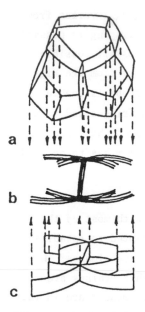

FIG. 2 Example of an erroneous reconstitution of the object from
002 LF: (a) Object, (b) 002 LF, and (c) ribbon-like erroneous
reconstitution. (From Ref. 6.)

B. Dark-Field Modes

1. *Principle*

By selecting only one hkl beam to form the image, the regions of the
object having scattered this beam appear bright on a dark field (DF).
A suitable small objective aperture is inserted in the back focal
plane of the objective lens (SAD pattern plane or Abbe plane) so as
to let through only the hkl beam.

Practically, the spherical aberration spoils the image, so that
a tilted dark field must be employed [7]. The aperture is set
paraxial and the scattered beam is brought paraxial by tilting the
incident beam.

a. *Crystalline Materials*. In the case of a three-dimensional
crystal, the reciprocal nodes are small (see appendix). The scattered
beams are thus sharp. Therefore the Bragg condition is very strict
and any tilting of the beam makes the hkl reflection disappear. A
goniometer stage must be employed [Fig. 3(a)] to first bring the hkl

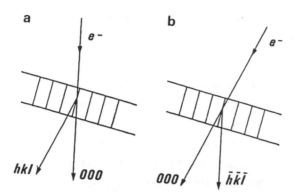

FIG. 3   Principle of tilted DF for a crystal.   (From Ref. 6.)

family of planes at the Bragg angle.   Then the beam is tilted [Fig.
3(b)] to bring $\bar{h}\bar{k}\bar{l}$ paraxial.   Thus hkl dims and disappears whereas
$\bar{h}\bar{k}\bar{l}$ attains its maximum intensity.

b. Disordered Materials.   As the crystal size decreases, the reciprocal nodes enlarge.   The Bragg condition is less and less strict so
that an increasing error relative to the Bragg angle becomes possible,
without losing too much intensity of the scattered beam.   It is thus
possible to avoid the use of a goniometer stage.   The tilted dark
field is obtained by simply tilting the incident beam so as to bring
a given hkl reflection into paraxial position.

As an example, a two-layer-thick carbon particle having a 1-nm
diameter (L_a) is supposed to be deposited edge-on on the supporting
film.   When the angle between the aromatic layers and the incident
beam reaches θ_{002} (i.e., a little less than 5×10^{-3} rad, since d_{002}
is larger here than 0.335 nm), the 002 beam will have its maximum
intensity.   The center of the 002 reciprocal node will be on the
Ewald sphere.   This node is in fact a spike 1 nm^{-1} in length (width
at half maximum of the node, that is, $1/L_a$).   The 002 beam will vanish
for a tilt α of the particle equal to $\pm d_{002}/2L_a$, that is, about 17-
18×10^{-2} rad (about $\pm 10°$).   A $2\theta_{002}$ tilting of the beam will be negligible relative to this value.   If the thickness of the particle
increases much (L_a = 10 nm), the maximum permissible tilt α before

the disappearance of the 002 beam becomes 10 times less. Then α
approaches the order of magnitude of 2θ.

2. *Resolution*

The contrast in a DF image is mainly due to the intensity of the hkl
beam. Since carbon specimens are thin and the atomic number Z is
small, the scattered beams are faint. It is therefore necessary to
work with a highly concentrated electron beam (cross-over). Correspondingly, the illumination is highly incoherent. In this case, the
resolution δ is given by the Abbe formula $\delta = 0.61\lambda/\alpha_0$, where α_0 is
the half opening of the aperture [8,9]. This opening must be main-
tained as large as possible to insure the maximum resolution, and
also must be chosen as small as possible to limit at the minimum the
portion of the scattered beam or beams allowed to contribute to the
image.

As an example, for an objective aperture of 2.4 nm^{-1} in diameter,
δ is a little larger than 0.6 nm. Such an aperture is too small to
ensure the optimum point-to-point resolution. The corresponding portion of the reciprocal space available for exploration is large near
the origin and decreases as the aperture is centered at increasingly
wider angles. However, it is always noticeable. This feature can be
expressed in terms of the range of d_{hkl} spacings corresponding to the
diameter of the aperture for different positions of centering. It can
be also expressed in terms of possible tilts α and β twists of the
object in the real space. The calculated data are given in Table 2.

Such an aperture allows one to easily separate each carbon-
scattered beam radially. It also allows one to collect any graphite
single-crystal hk0 reflection without being disturbed by the five
other ones. To improve the resolution down to 0.35 nm, the aperture
has to be increased to 3.5 nm^{-1} in diameter. For that size 002 and
100 cannot be separated from each other. However, the six 100 reflections are kept separated up to an aperture diameter of 4.7 nm^{-1}.

Two remarks are necessary at this point:

1. If the regions issuing a given hkl beam are small, the beam
    is broad and faint. Nevertheless since the intensity con-

TABLE 2 Possible Latitude for Tilting and Twisting in the Case of an Objective Aperture 2.4 nm^{-1} in Diameter Centered on Carbon Beams

Carbon scattered beams	d_{hkl} range (nm)	Tilt $\alpha°/2$[a]	Twist $\beta°/2$
002	0.550-0.238	± 44	± 22
10	0.288-0.170	± 37	± 14
11	0.144-0.107	± 29	± 8

[a]α is calculated from cos α = 1/(1 + rd_{hkl}) where r is the radius of the aperture.

        centrates in a small area of the image, the elementary scattering domain is bright enough to be seen.

2. If the regions issuing the hkl beam are smaller than the resolution δ they are imaged as bright domains, the size of which is meaningless since it is equal to δ.

3. Percentage of Scattering Domains Imaged

If the scattering domains are distributed at random in the specimen, the selected area electron diffraction (SAD) pattern is a Debye-Scherrer powder pattern. Knowing the size of the objective aperture used, one can calculate the percentage of scattering domains imaged [10]. If t is the thickness of the scattering domain along the incident beam direction, the solid angle under which the hkl reciprocal nodes are seen is Ω = 2πd_{hkl}/t. The percentage of scattering domains is p = 2πd_{hkl}/4πt, that is, d_{hkl}/2t. If m is the multiplicity factor of the hkl planes and f the portion of the Debye-Scherrer ring selected, then p = md_{hkl}f/2t. Considering again the example of a 002 dark field of a 1-nm carbon particle, p = 8.5%. The specimen thickness for which the image is entirely filled with the images of the 002 scattering domains can also be calculated. If T is the specimen thickness, the number of scattering domains per unit of surface of the specimen is T/t^3. The number of scattering domains fulfilling the Bragg condition is fmd_{hkl}T/2t^4, since the domain surface is t^2. As a result,

$$T^2 = \frac{2t^2}{fmd_{hkl}}$$

In the above example T = 22 nm. If the specimen thickness is smaller than T, the smaller it is, the smaller are the chances of superimposition of the individual images. In any case, the chances are much smaller than for axial illumination lattice fringes (f = 1).

It must be noticed that in the case of a preferred orientation within the specimen (parallel or fibrous), and should its extent reach the size of the intermediate aperture used for selected area electron diffraction (~1 μm), the above formula does not apply. It is then necessary to recalculate Ω for each particular case [11]. Moreover, if the specimen is porous, the porosity has to be taken into account since it is an advantage for decreasing the possible image super-impositions.

4. Imaging of Very Small Aromatic Ring Structures

a. Units at Random within the Specimen. It is commonly admitted that most of the very-low-temperature carbonaceous materials, or immature natural carbonaceous matter, contain small planar aromatic ring structures, more or less piled up in stacks connected to each other by nonaromatic functional groups [12,13]. Such aromatic ring structures can be considered as small graphitic layers, hexagonal in symmetry. Whatever the orientation of a single layer, its atom rows scatter electrons, and 12 beams (six 10 and six 11) are emitted (see appendix). If two aromatic structures happen to be piled up in parallel, additional 00ℓ scattered beams will appear. As an example, Fig. 4(a) represents single and piled up ring structures distributed at random in the specimen [14]. The nonaromatic cement is also shown. The model represents the organic fraction of a low-rank parent rock of oil (kerogen). Some of the aromatic ring structures piled up by two are shaded and numbered 1, 2, and 3. They are parallel to the incident beam and thus approximately fulfill the 002 Bragg condition. A single structure is numbered 4. It yields the 10 and 11 beams having the same number in the SAD pattern [inset in Fig. 4(a)]. Stacks 1, 2, and 3 would yield pairs of 002 beams numbered 1, 2, and

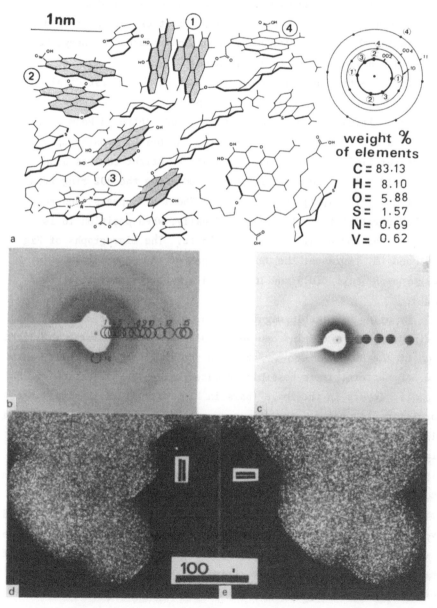

FIG. 4 (a) Model of low-temperature carbonaceous material. Inset: SAD pattern sketch. (b) Displacement of the objective aperture. (c) Actual image. (d and e) Orthogonal 002 DF (positions 5 and 16). The aromatic layers are represented by a double bar. (From Refs. 14 and 3.)

3 in the inset. In the case of random single layers alone, only 10
and 11 asymmetrical bands are present. If stacks are contained in
the sample, the overall random orientation produces 00ℓ Debye-Scherrer
rings in addition to the 10 and 11 bands.

The objective aperture [Fig. 4(b) and (c)] is first displaced
radially relative to the SAD pattern (positions 1-15). When it
reaches position 5 [corresponding to the 002 scattered beam of the
stacks parallel to 1 light up altogether as bright domains upon a
stacks parallel to 1 light up altogether as bright comains upon a
dark field. When the aperture is rotated along the 002 ring from
position 5 [Fig. 4(b)] to position 16, BSU 1 disappears, and then
all BSU parallel to 3 illuminate. Finally, when position 16 is
reached, the BSU parallel to 2 illuminate. The micrographs of Fig.
4(d) and (e) represent the 002 DF images of the same particle in
positions 5 [Fig. 4(d)] and 16 [Fig. 4(e)] corresponding to BSU 1
and 2.

By using a suitable magnification, a *histogram of sizes* of the
individual bright domains can be established [15]. Knowing the direction of the 002 scattering vector and the image rotation relative to
the SAD pattern, it is possible to materialize the projection of the
aromatic layers in the image (bars in Fig. 4). It is thus possible
to distinguish between the diameter and the thickness of the BSU. In
all samples studied, BSU were isometric and always less than 1 nm in
size. Figure 5 shows an example of histograms established for two
pitches. Figure 5(a) corresponds to a coal-tar pitch. The histogram
is practically symmetrical with a mean size of 0.66 nm. Figure 5(b)
corresponds to a petroleum pitch. Here, the measurement is disturbed
by the resolution limit. The majority of BSU is artificially gathered
into the first class, which approximately corresponds to the value of
the resolution δ. The BSU are therefore smaller than those of the
coal tar pitch, although to an unknown extent. The possible models
of aromatic ring structure fitting 002 DF images are given in Fig.
5(c). The best agreement is found for the smallest molecules such
as corenene.

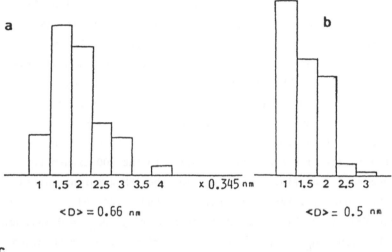

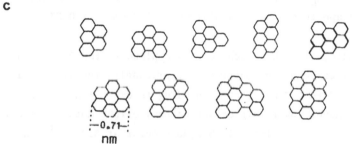

FIG. 5 Histograms of size of 002 scattering domains (BSU). (a) Coal tar pitch. (b) Petroleum pitch. (From Ref. 15.) (c) Models of BSU. (From Ref. 14.)

More than 400 products have been studied up to now, among them kerogens [14], coals [16,17], pitches [15,18], petroleum derivatives [19-21], anthracene-based carbons (AC), and saccharose-based carbons (SC) [3,22,23]. All of them contain BSU at random in the bulk. None of the latter are larger than 1 nm. Correspondingly, the stacked molecules are piled up by two or three at most.

As the heat treatment temperature (HTT) increases, such units do not grow in thickness up to about 1000°C (heating rate 4° min^{-1}) and in diameter up to 1500°C [3,23].

Some idea of the *interlayer spacing* of a BSU is given by a radial exploration of the reciprocal space in the vicinities of

positions 5 or 16, using a particularly small aperture (~ 1 nm^{-1}).
This reduces the resolution to more than 12 nm [24], but it also
decreases the range of interplanar spacing selected by the aperture,
i.e., the error in the numerical value of d. The experimental procedure is as follows. The beam is progressively tilted toward higher
angles, whereas the 002 DF image is observed on the fluorescent
screen. Moving the aperture is continued until the first bright
domains appear. Then, going back to the SAD adjustment of the
lenses, the shadow of the aperture relative to the 000 transmitted
beam is recorded [see Fig. 4(c)]. Knowing the TEM apparent camera
length, the range of interlayer spacings is known [24-26]. The error
calculated from a graphite 002 Debye-Scherrer ring ($d_{002} = 0.335_4$ nm)
is given by $0.29 < d_{002} < 0.39$ nm. The experimental error obtained
by reducing the aperture size to a minimum of 1.2 nm^{-1} is $0.33 <$
$d_{002} < 0.35$ nm.

     As an example, if such a technique is applied to the BSU of an
immature natural carbonaceous material, surprisingly, the first bright
domains appear before 1.3 nm^{-1}, corresponding to $d > 0.8$ nm. When
the aperture is moved toward higher angles many bright dots disappear
while others show up. The last visible dots occur at 3.2 nm^{-1}, in
the vicinity of the spacing of a turbostratic carbon. The d spacing
of the maximum number of bright domains ranges between 0.4 and 0.5
nm. When such a sample is heat-treated under an inert gas flow, the
interlayer spacing spreading decreases and the average values progressively shift toward the lowest value. No more spreading is observed after heat treatment at 1000°C at a 4° min^{-1} heating rate.

     In the case of such small units, hk dark fields corresponding
to positions 8 and 12 in Fig. 4(b) do not give many additional data
except the certainty to observe turbostratic aromatic ring structures,
since bright domains are observed only for these positions.

b. Occurrence of Local Preferred Orientation. If some BSU associate
edge to edge and face to face with their aromatic layers in parallel,
the 002 scattered beams reach the same point of the SAD pattern and
thus pass altogether through the objective aperture. The oriented

area is imaged as a cluster of bright domains [Fig. 6(a)]. Each
domain images a single BSU, and the extent of the cluster [circled
in Fig. 6(a)] is that of the local molecular orientation (LMO). The
dark areas in the image of Fig. 6(a) correspond to other LMO, either
twisted or tilted too much for letting their 002 beams pass through
the aperture. LMO will not vanish in the DF image as long as the 002
beams do not escape the aperture. Figure 1(c) (see also Fig. 7) shows
that the possible deviation from a planar configuration is given first
by the angle β (Table 2), i.e., by the angular portion of the 002 ring
intersected by the aperture. The limiting tilt α is not that given in
Table 2, since it is not fixed by the aperture opening. It is only
related to the maximum permissible error for fulfilling the Bragg
condition. As demonstrated previously, a BSU 1 nm in diameter toler-
ates $\alpha = \pm 10°$ before its images disappear in 002 DF. The LMO follows
the same rule.

    To image all the LMO it is necessary first to rotate the aperture
along the 002 ring [Fig. 6(a) and (b)], then to tilt the sample using
a goniometer stage [6,27]. A *three-dimensional reconstruction of
microtexture is thus possible* [14,27-30], even if the elemental units
are very small (<1 nm). The better organized the product, the easier
it is to obtain these data.

    The parallel orientation of two adjacent BSU in itself is gener-
ally not perfect and suffers deviations by both a tilt α and a twist
β (Fig. 7). For a perfect orientation and a flat planar LMO the common 002 reflection of all BSU in the pattern lies on the normal to the
carbon layers. If there is no preferred orientation the normals are
uniformly distributed on a sphere. The 002 reflections are thus also
uniformly distributed on a sphere (powder sphere). If a certain
amount of misorientation is due to either twist β or tilt α of the BSU
relative to the mean orientation plane (Fig. 7), the density of normals
decreases inside a cone limiting a portion of this sphere. The maximum
density occupies the axis of the cone. The curves representing the
density distribution of the normals are usually determined along two
perpendicular arcs, respectively corresponding to α and β.

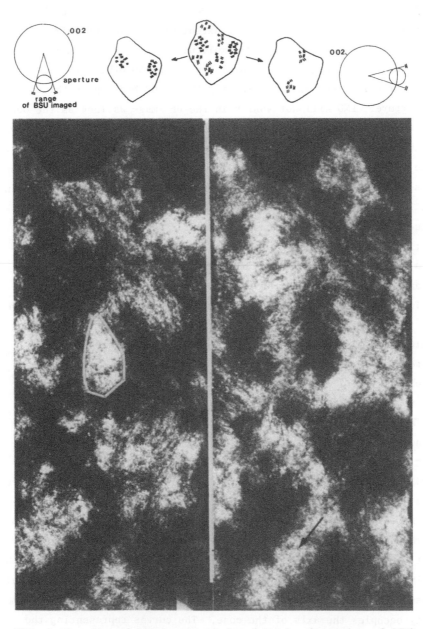

FIG. 6 Imaging of LMO in 002 DF (preferred orientation of BSU). At the top, sketches; below, images for two orthogonal positions of the aperture. (From Ref. 6.)

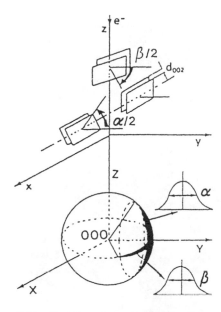

FIG. 7 Misorientation of BSU: twist β, tilt α. The term β is limited by the aperture opening, and α by the maximum permissible tilt for which the 002 reciprocal node still touches the Edwald sphere.

Figures 8 and 9 illustrate in more detail how the *three-dimensional arrangement of BSU* can be deduced from *azimuthal explora-tion* by rotating the aperture relative to the 002 ring (Fig. 8), that is, rotating BSU around OZ in Fig. 7, and from *tilting the sample* (Fig. 9). The tilting axis was chosen so as to let the aromatic layers fulfil the 002 Bragg condition during the whole experiment (rotation axis OY perpendicular to the layers, i.e., corresponding to the scattering vector $\vec{s}_{002}$ reaching the aperture center). Another direction would lead to the extinction of the bright area. Let us consider now a given cluster of bright dots whose thickness is h (unknown) and width l_0 [measured on Fig. 9(a)] for the initial position of the sample [Fig. 10(a)]. After a known tilting i, l_0 becomes $l_{(i)}$ so that $l_{(i)} < l_0$ with $l_{(i)} = l_0 \cos i - h \sin i$ [Fig. 10(b)].

For most low-temperature chars h is usually about equal to l_0. The LMO is isometric.

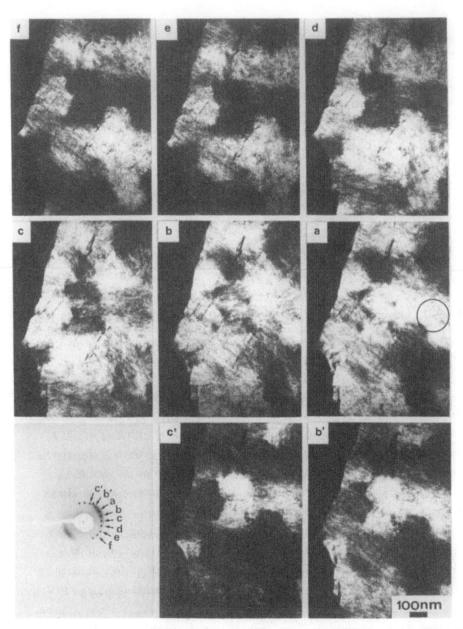

FIG. 8 Azimuthal exploration of the 002 ring starting from (a) 002 DF;
inset, SAD pattern showing the aperture positions. The misorientation
β of the BSU can be seen from the circled area in (a). (From Ref. 29.)

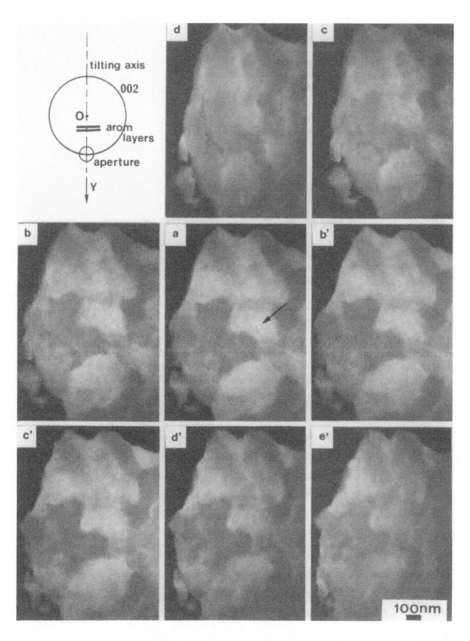

FIG. 9 Specimen positive tilting by 12° steps from a to c; negative tilting from a to e'. The tilt axis is sketched relative to the aperture and to the layers. (From Ref. 29.)

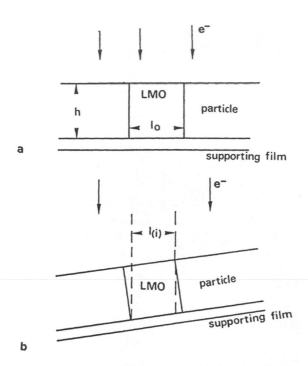

FIG. 10 Measurement of the thickness of LMO with a goniometer
stage: (a) before tilting, and (b) after tilting. (From Ref. 14.)

The terms α *and* β*, the tilt and twist between BSU inside a given
LMO, can be measured* through photometric recordings of various DF
images. To measure β the aperture is moved to and from the position
corresponding to the maximum brightness of the center of a given LMO
[for instance, the circled area in Fig. 8(a)]. For example, the BSU
are rotated around OZ in Fig. 7. The intensity is then recorded
during the aperture displacement. To measure α, a goniometer stage
is used and the tilting axis is chosen parallel to the aromatic layers,
normal to the scattering vector reaching the aperture center. It is
OX in Fig. 7. In this case, the LMO vanished as tilting increases.
The recording of the changes in brightness yields the curve of α.

5. *Imaging of Relatively Large Aromatic Layers*
Whatever its size, an aromatic layer stack is more or less distorted
because it is thin. Nevertheless, the larger the diameter L_a and the

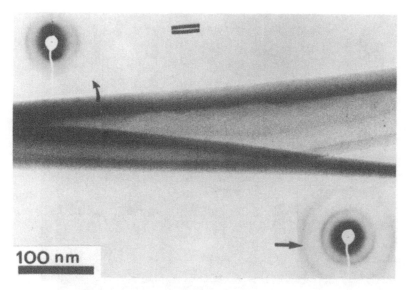

FIG. 11 Fold in a thin carbon film (as deposited), BF. Top inset,
SAD pattern of the fold; bottom inset, SAD pattern of the flat area.

thickness L_c the smallest the 00ℓ nodes $(1/L_a, 1/L_c)$ and the thinner
the hk lines $(1/L_a)$. The DF images will therefore be highly modified.

Heat-treated thin carbon films will be presented as an example
[31,32]. They are made of aromatic layers parallel to the film plane,
so that when such a film breaks it folds, thus bringing the aromatic
layers edge on (Fig. 11).

a. *002 DF*. At a low HTT, the fold appears full of bright dots
imaging the individual BSU [Fig. 12(a)]. Because a large tilt ($\pm$ 10°)
is permissible before 002 vanishes, the fold contour is not sharp. At
a high HTT L_c and L_a increase so that the 002 node ceases very rapidly
to touch the Ewald sphere because $1/L_a$ is small. Even a very weak
tilting distortion makes 002 disappear. The fold contour is thus
sharp [Fig. 12(b)]. The loci of the regions fulfilling the 002 Bragg
condition are now numerous, thin, bright, and distorted lines or Bragg
fringes [Fig. 12(b), arrows].

At this point it is interesting to compare 002 DF data to 002 LF
[Fig. 12(c) and (d)]. Individual BSU are imaged in Fig. 12(c) as two

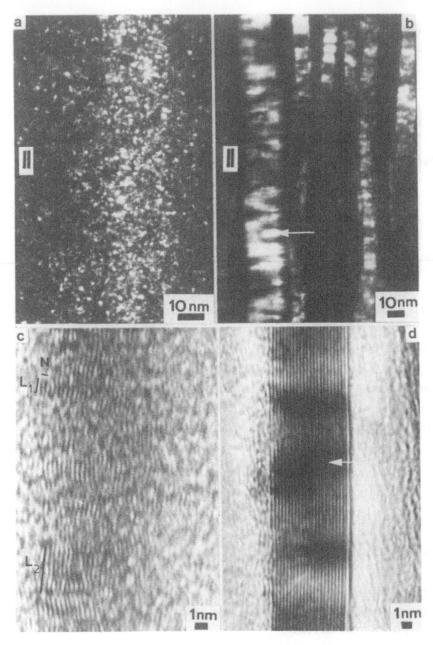

FIG. 12 Folds in carbon film. (a,b) 002 DF, with (a) low HTT and
(b) high HTT (arrow, Bragg fringes). (c,d) 002 LF, with (c) low HTT
and (d) high HTT (arrow, Bragg fringes). (From Ref. 31.)

to three (N) parallel fringes less than 1 nm (L$_1$) in length (low HTT).
Thick stacks (N $\approx$ 30) of long and relatively perfect fringes are
imaged in Fig. 12(d) (high HTT), showing dark Bragg fringes (arrows).

b. *hk Dark Fields*. As soon as the diameter of the layers becomes
noticeable, whether turbostratic or not, numerous hk DF data can be
added to 002 data [3,6,30-33].

As single aromatic layers are superimposed with a slight rotation
(turbostratic structure), hk rotational moiré fringes are produced
(Fig. 13). Their spacing D is related to γ (the rotation angle) and
d$_{hk}$ (the distance between the carbon atom rows of each single layer)
by the formula

$$D = \frac{d_{hk}}{\gamma}$$

The direction of the moiré fringes is parallel to the bisector of
the angle γ, roughly perpendicular to the scattering lattice rows.
This phenomenon is due to the fact that two hk beams are close enough
to each other to interfere. The moiré fringes will thus be parallel
to the scattering vector reaching the objective aperture center.

In the case of turbostratic material there is a large number of
superimposed layers whose scattered beams are close to each other and
pass through the aperture. However, the contours of these pairs of
layers are not coincident, neither along the thickness of the specimen
nor in the image, which is a bidimensional projection. Correspondingly, the moiré fringes do not occupy localized areas in the image.
They cover the whole area of the specimen observed as "dotted"
fringes [Fig. 14(a)] whose general direction is fixed by the aperture

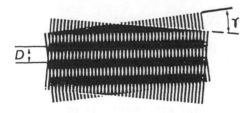

FIG. 13 Rotational moiré patterns.

FIG. 14 Moiré fringes in a carbon film. (a) HTT 1760°C, 11 DF,
turbostratic moiré fringes. (From Ref. 16.) (b) HTT 2340°C,
110 DF, rotational moirés. (From Ref. 31.)

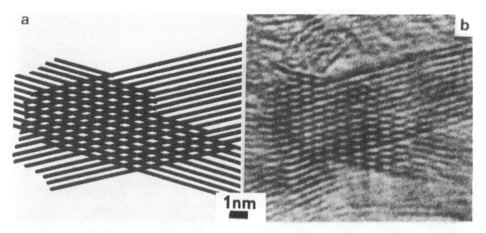

FIG. 15   002 LF, rotational moirés: (a) sketch and (b) image.
(From Ref. 34.)

position relative to the SAD pattern, that is, parallel to $\vec{s}_{11}$.   The
"turbostratic" moiré fringes thus have a very characteristic common
feature.

In the same manner, two graphite crystallites superimposed at
random give rotational moirés.   In this case, the fringes occur in a
localized area representing the projection of the common part of the
two crystals.   The boundaries between two grains are thus clearly
marked [Fig. 14(b)].   The crystals not fulfilling the 110 Bragg condition remain dark.   The ones not superimposed on others but more or
less fulfilling the 110 Bragg condition appear homogeneously grey.
The arrow in Fig. 14(b) gives an example of a small crystal superimposed on only one part of a larger one.   The double arrow is an
example of superimposition of more than two crystals.

The data obtained from such micrographs are the *size* (diameter
La_{11Cr}) and the *shape* of the crystallites, whereas for *turbostratic*
materials, one obtains only the mean length of the fringes which is
only *indirectly* related to the layer *diameter* (La_{11TF}).

The same kind of rotational moiré patterns can be observed in
002 LF [compare the sketch of Fig. 15(a) to the micrograph of Fig.
15(b)].

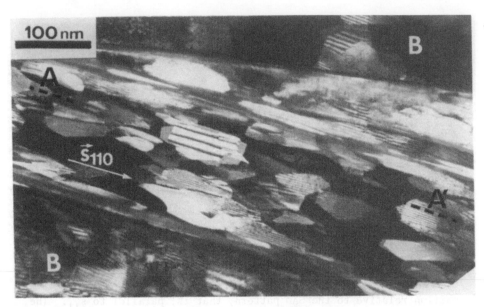

FIG. 16   Carbon film (HTT 2340°C) of Fig. 14(b) folded along AA',
110 DF.   (From Ref. 31.)

One of the main interests of the hk moiré images is to give a
reliable representation of both the orientation and the distortions
of the specimen along the normal to the observation plane.

1. *Inclined specimen* (crystallites smaller than the folded
    area).   Comparing Fig. 16 to Fig. 14(b) shows the effect of
    the inclination of the specimen due to a fold in the thin
    film represented flat in Fig. 14(b).   Since the fold axis
    AA' is parallel to the 110 scattering vector reaching the
    aperture center, the 110 planes are perpendicular to AA'.
    Hence they still fulfill approximately the 110 Bragg condi-
    tion in the whole micrograph even in the flat part of the
    film B.   The crystallites are no longer hexagonal but elon-
    gated along AA' and the periods of the moiré fringes are
    still visible.

2. *Distorted specimen.*   If a given turbostratic carbon layer
    stack is now larger than the fold itself, that is, distorted
    (Fig. 17), two possibilities arise.   The first [Fig. 18 and

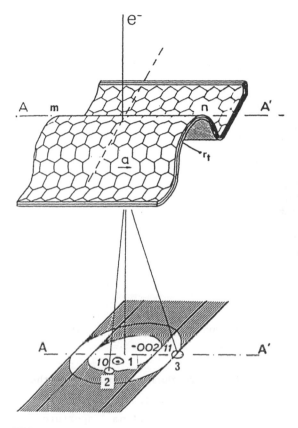

FIG. 17 Distorted turbostratic stack and SAD pattern. Three positions of the objective aperture are represented: 1 on 002, 2 on 10, and 3 on 11.

19(d)] corresponds to the same conditions as in Fig. 16, that is, the fold axis AA' is parallel to the operative scattering vector $\vec{s}$ (position 3 in Fig. 17). The stack thus always yields a scattered beam passing through the aperture. The moiré fringes are parallel to AA'. The moiré period D is constant in the F_1, F_2, F_3 flat parts (Fig. 18). It decreases in the two oblique O_1, O_2 areas. The second corresponds to an operative scattering vector $\vec{s}_{hk}$ chosen normal to AA' [Fig. 19(c), position 2 in Fig. 17]. The moiré fringes are perpendicular to AA'. The hk beams issued from the stack escape the aperture opening beyond a certain tilt fixed by

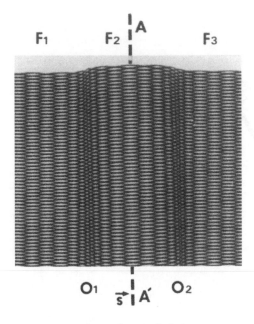

FIG. 18 Model of moiré fringes corresponding to position 3 in
Fig. 17.

the aperture size. The fringes are thus interrupted in the
oblique parts of the fold. As an example, the maximum permissible tilt is ±37° if $\vec{s}$ is 10, for a 2.4 nm^{-1} aperture
(Table 2). The image is then made of ribbon-like rotational
moiré areas. If r_t is the radius of curvature of the fold,
one can calculate the possible 002 (position 1 in Fig. 17),
10 (position 2), and 11 (position 3) dark-field images,
taking into account the size of the objective aperture
opening given in Table 2. For the 2.4 nm^{-1} aperture, Fig.
19 sketches the images and gives the data obtainable. The
top of Fig. 19(a) represents on the left the profile of a
smoothly distorted layer stack and on the right a strongly
folded one. Figure 19(b) represents 002 DF, 19(c) 10 DF,
and 19(d) 11 DF. (For imaging the same stack the aperture
must be successively centered as indicated in Fig. 17.) *By
combining all the data given by all possible hkl DF images,*

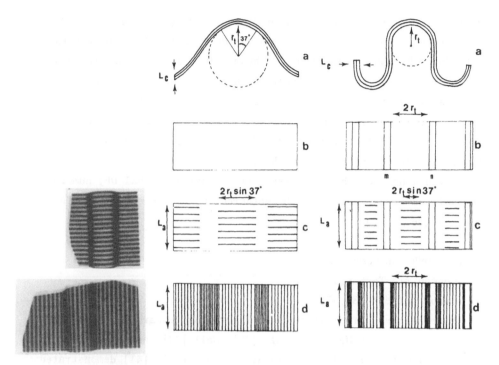

FIG. 19 Increasingly distorted stacks: (a) profile, (b) 002 DF
(position 1 of Fig. 17), (c) 10 DF (position 2), and (d) 11 DF
(position 3). (From Ref. 11.)

> *it is possible to obtain a complete three-dimensional recon-*
> *struction of microtexture, whatever the complexity of the*
> *specimen.*[*]

C. Conclusion

If the instrument setting can be known with enough accuracy, 002 LF
images can be used concurrently with 002 DF. Despite the fact that
both are based on the Fourier transform of the same 002 scattered
beam they overlap but they are partly complementary. The 002 LF
correspond to a two-beam interference, modified by some additional

---

[*]Even a specimen distorted in many directions and thus giving a complete SAD powder pattern can be described by dividing the pattern in
successive areas fulfilling the relative locations 1, 2, 3 of Fig. 17.
Series of sets of three micrographs are then recorded for each area
and the data suitably associated (see for example Ref. 35 and Section
IIIA1b).

diffraction contrast due to higher order of 00ℓ (occurrence of 004
Bragg fringes for example), whereas 002 DF correspond to the contrast
introduced by the intensity of one beam only. Up to a certain point,
002 DF images can be used to control lattice-fringe data, since the
changes of superimposition of BSU images are smaller (see Section
IIB3). Reciprocally, since individual fringes mark single aromatic
layers, additional data are obtained from LF such as the number N of
layers per stack and their length L_1.

By adding then the data issued from other hk or hkl DF, numerous
valuable numerical values can be gathered, which are summarized in
Table 3 (see p. 79).

III.   SPHERICAL CONFIGURATIONS

A.   Concentric Textures

1.   *Carbon Blacks* [36,37]

a. *Microtexture.* The spherical configuration of carbon blacks was
emphasized first by Hofmann et al. [38], Hall [39], and Boehm [40],
using DF TEM techniques. Then Kasatotchkine et al. [41] demonstrated
this texture by SAD. They used a high-voltage TEM (400 kV) and performed patterns around a single particle of thermal carbon black heat-
treated at 3200°C. Later, indirect evidence of the concentric texture
was obtained by observation in bright field of oxidized carbon blacks
by Donnet et al. [42-44] and also Heckmann et al. [45]. Other DF data
were obtained by Rudee [46].

An example of 002 DF imaging is given in Fig. 20(a,b) and (c,d).
Small bright domains less than 1 nm in size are gathered into two
sectors. The angular opening of each sector is equal to that of the
objective aperture chosen. Its angle is a function of the opening
diameter. The sectors turn clockwise or counterclockwise in the same
amount as the aperture is displaced along the 002 ring. Since these
bright sectors are always obtained for any particles, all of the
latter have necessarily a rotational symmetry (spherical or cylindrical). The 002 lattice-fringe images were finally obtained by Heidenreich et al. [1], then by Marsh et al. [47], as shown in Fig. 20(f).

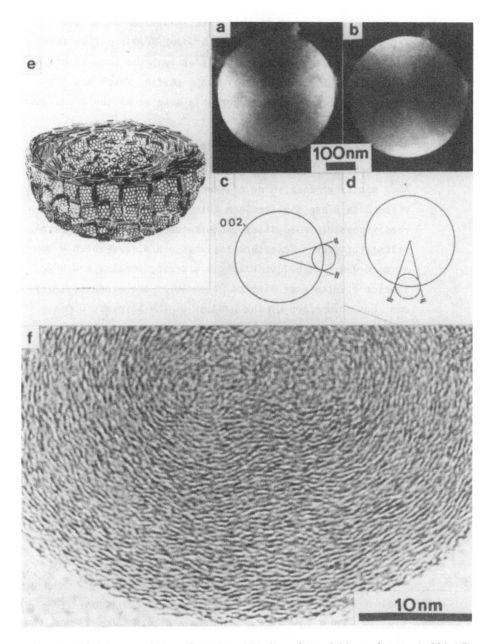

FIG. 20 Single particle of carbon black: (a and b) orthogonal 002 DF, (c and d) sketches of the aperture position, (e) model [1], and (f) 002 LF.

It is interesting to stop at this point and to consider the
above data. All authors agree in giving a model of onion-like texture [Fig. 20(e)] for carbon blacks. In addition, the lack of bright
domains or of fringes in the center of the particle, which was generally observed, was interpreted either as a hole or as the occurrence
of amorphous carbon.

Let us discuss these two assumptions:

1. The spherical configuration was postulated only intuitively.
   As a matter of fact, none of the techniques employed could
   yield data along the specimen thickness. Hence it was not
   really possible to distinguish between spherical and cylindrical textures. Nevertheless, what had already been demonstrated was that 002 DF bright scattering domains and 002
   lattice fringes were limited not only by the aromatic layer
   end, but more often by the bending of the layers. Consequently, nothing in the previous works proved that carbon
   blacks were not platelets with cylindrical texture. The
   only way to be sure of the onion-like texture was to observe
   other hk DF images. This was done by Oberlin and Terrière
   [48] in combining 002, 10, and 11 DF imaging by a radial
   exploration of the SAD pattern (see Fig. 4). Figure 21
   shows the complete sequence obtained from the bright-field
   image [Fig. 21(a)]. The 002 image of Fig. 21(b) shows again
   the two bright sectors corresponding to the imaging of C and
   D regions of the model [Fig. 21(e)]. If the single particle
   is spherical, its diffraction pattern will be a Debye-Scherrer
   pattern. Such a pattern derives from the pattern of Fig. 17
   by a rotation around the incident beam. It is therefore
   possible to consider only two orthogonal bands AOB and COD in
   the model of Fig. 21(e). Each of them can be approximated by
   Fig. 17 with a suitable orientation. Their SAD patterns can
   also be represented by Fig. 17 with a suitable orientation.
   Figure 22(a) shows the superimposition of the pattern of AOB
   represented in dashed lines and COD represented in heavy

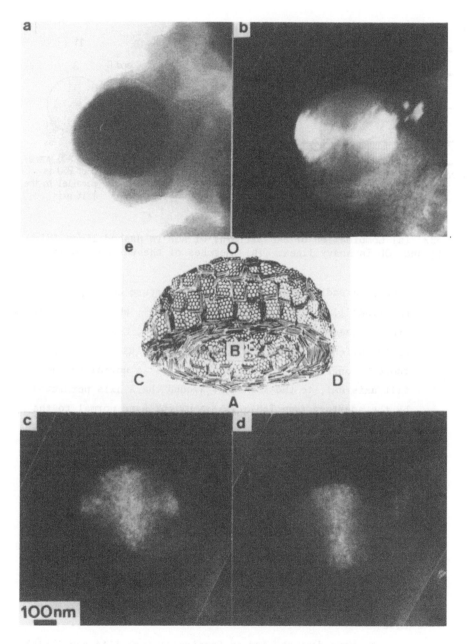

FIG. 21 Carbon black, DF radial exploration. (a) BF, (b) 002 DF
[position 1 of Fig. 22(a)], (c) 10 DF (position 2), (d) 11 DF
(position 3), and (e) model (in b, C and D are bright; in c, C and
D and AOB are bright; in d, AOB is bright). (From Ref. 48.)

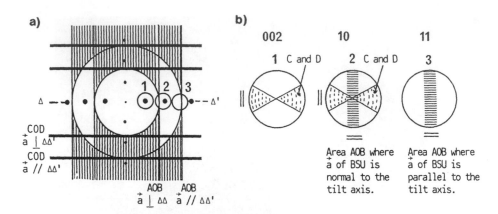

FIG. 22   (a) Combination of SAD patterns of AOB in dashed lines [Fig. 21(e)] and COD in heavy lines.   (b) Sketches of the various images.

lines.   A radial exploration by the aperture corresponds to positions 1, 2, and 3 on $\Delta\Delta'$ in Fig. 22(a), whereas the images are sketched in Fig. 22(b).   In position 1, C and D are imaged as two bright sectors.   In position 2, the BSU imaged are those belonging to area AOB whose a axis is normal to the tilt axis $\Delta\Delta'$, to the $\pm37°$ tilt around the $\vec{a}$ axis permitted by the aperture (see Table 2).   A bright band normal to the sectors will thus appear.   Because 004 of the C and D areas pass through the aperture, faint sectors are visible in the same position as for 002 DF.   In position 3, the BSU imaged are those belonging to area AOB whose $\vec{a}$ axis is parallel to the tilt axis $\Delta\Delta'$ (see Table 2).   A bright band thus appears in the same position as in position 2.   Since 006 is not intersected by the aperture, no sectors appear.

2.   Figure 23(a) [see also Fig. 20(f)] is an example of the "hole" commonly encountered in carbon blacks.   Is it really a hole, an amorphous or nearly amorphous area, or something else?

   We have demonstrated in Section II (see IIA4 and IIB1b) that the maximum permissible tilt for a single BSU (<1 nm in size) to still fulfill the 002 Bragg condition was $\pm10°$.

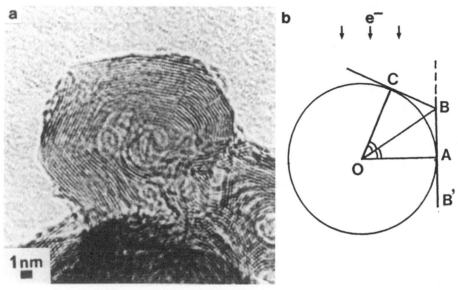

FIG. 23   (a) Single particle of carbon black showing a "hole" in the
center.   (b) Calculation of the radius of the empty core.

Ideally, there is no BSU visible beyond a sector of ±10°
around the equator of a spherical particle. Correspondingly,
because the radius of curvature of the concentric sheets of
BSU decreases from surface to core, the chances to have BSU
imaged either in 002 DF or 002 LF are null at the center,
even if no disturbance occurs in texture. Let us consider
two adjacent BSU tangential to a circle near the center of
the sphere [Fig. 23(b)]. OB is bisector of AOC. The angle
CBO is thus equal to OBA. If AB of BSU 1 fulfills the 002
Bragg condition, BOA is 10° at the maximum. BC of BSU 2
cannot be imaged. The radius of the circle OA is given by

$$\frac{AB}{OA} = \tan 10° \qquad \text{that is,} \qquad OA = \frac{0.5}{0.176} \approx 2.8 \text{ nm}$$

Theoretically, whatever the diameter of the sphere, there
is no imaging possible inside a circle 5.6 nm in diameter.

b. Various Types of Carbon Blacks. i. Spherical or subspherical single particle [1,36-37,42-47]. The particles with a recognizable spherical shape always have a concentric texture: perfect in single spheres, and more or less deformed when single spheres associate together. Figure 20 shows a single sphere whereas Fig. 24 shows two possible types of aggregates: linear [Fig. 24(a)] or three-dimensional [Fig. 24(b)]. Such aggregates are produced by collapsed nuclei surrounded by layers held in common [Fig. 24(c)]. The concentric texture is then distorted and reduced to the external shell.

When these carbon blacks are prepared or heat-treated at higher temperatures, they become polyhedral, though initial single spheres remain recognizable (Fig. 25). The comparison between Figs. 25 and 20(g) allows us to understand change in shape from sphere to polyhedral. BSU with tilt and twist boundaries form the low-temperature spherical particle [Fig. 20(g)]. After heat treatment above 2000°C, defects frozen in at the boundaries are wiped out. Larger stiff and flat layer stacks are thus produced (Fig. 25). Necessarily, this causes the sphere to become polyhedral. Correspondingly, graphitization begins to occur. The very interesting feature [49-51] is that the size of flat regions never surpasses about one-third of the particle diameter. Correspondingly, the final degree of graphitization reached increases with the diameter of the initial spherical particle [51-55]. This means that the maximum permissible diameter of the flat layers is the leading parameter of graphitization. Similar considerations will be developed in the next section concerning other materials.

Among the examples given in the figures of this section, all electrically conductive carbon blacks, which are always very small, are nongraphitizing carbons [5], whereas thermal ones are partially graphitizing. Though nongraphitizing, some of the conductive carbon blacks show a noticeable decrease of d_{002} when heat-treated (Ensagri and Y50B in Fig. 26).

ii. Carbon blacks with statistical spherical symmetry (see also Section IIIC) [5]. The electrically conductive carbon blacks mentioned above are interesting for their applications in battery manu-

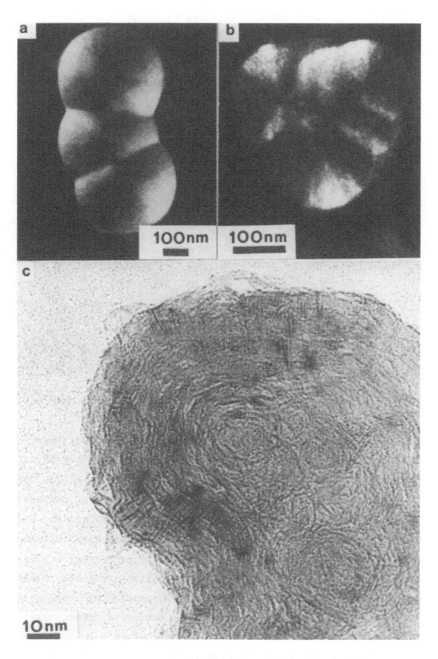

FIG. 24 Carbon blacks, 002 DF and LF: (a) chains, (b) aggregates, and (c) multiple nuclei.

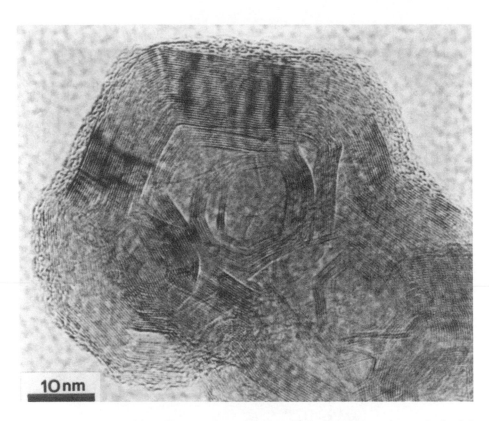

FIG. 25   Polyhedral carbon blacks (HTT 2800°C), 002 LF.   (From Ref. 5.)

facturing (Leclanché batteries).   For having good qualities, these
carbon blacks should exhibit large specific surfaces and provide a
small internal resistance to the battery.   The best material (S.70)
[56] has a very special microtexture.   Upon heat treatment, it does
not change either crystallographically (Fig. 26) or texturally.   No
spherical individual particles can be recognized, but only distorted
sheets of stiff and perfect carbon layers showing acute angles (Fig.
27).   Figure 27(a) does not permit us to establish a reliable three-
dimensional model because it might suggest very different models, such
as ribbon-like stacks or crumpled continuous sheets of paper.   To
avoid an erroneous model, the real configuration has to be established
by using hk DF, such as 11 DF for example [Fig. 27(b)].   Let us consider Fig. 19 in Section IIB5b.   The highly distorted sheet repre-

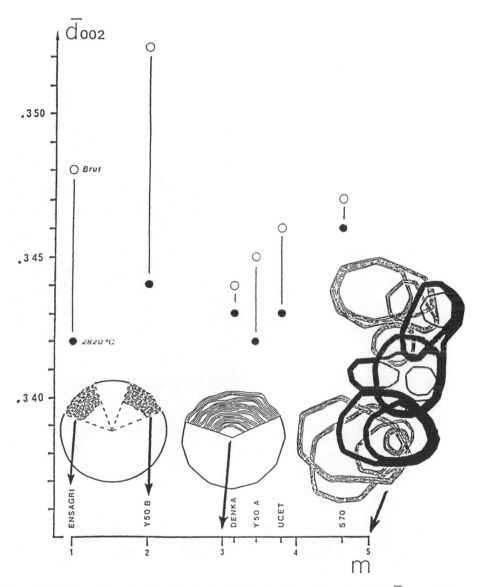

FIG. 26 Electrically conductive carbon blacks: decrease of $\overline{d}_{002}$ after heat treatment at 2800°C. Sketches of the corresponding types 1-5. (From Ref. 5.)

sented on the right yields a very typical 11 image (image d). The main feature of the image is the pair of bright diffuse lines framing a black band, occupying the m and n areas of the image folded zones

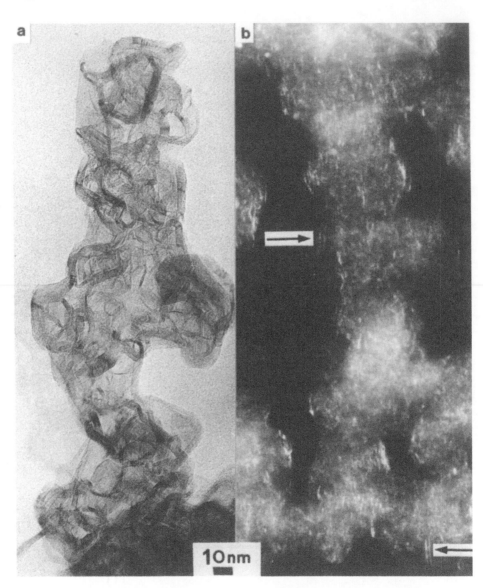

FIG. 27 Carbon black with statistical spherical symmetry (S.70).
(a) BF and (b) 11 DF. (From Ref. 5.)

(appearing bright in 002 DF). For such small carbon blacks, the rotational moirés are nearly indistinct. However, a noticeable amount of such pairs (arrows) occurs in Fig. 27(b) at the expected place, replacing the bright 002 bands. This figure thus demonstrates unambiguously that S.70 is made of distorted continuous sheets and not of ribbons. The same kinds of experiments were performed on other carbons and gave the same results (see next section). In materials such as S.70 the symmetry of the texture remains statistically spherical, since the distorted sheets are entangled along all possible directions.

[*Note*: In conductive carbon blacks all intermediates are known between Ensagri (spherical symmetry) and S.70 (statistically spherical symmetry), mostly belonging to five types sketched in Fig. 26. The abscissa m in Fig. 26 represents the arithmetic mean value of frequency histograms of the types present in a given black sample (most of them are heterogeneous). From Ensagri to S.70, the specific surface increases from 40 to 120 m^2/g, whereas the internal resistance of the battery decreases from more than 0.40 Ω down to 0.25 Ω. S.70 is thus much better than Ensagri for making batteries. In fact, it is the way the aromatic layer grows that induces one or another type of microtexture. The carbon blacks having real spherical particles (Ensagri, Y50B) grow radially by deposition of BSU upon nuclei, either single or multiple. At the opposite extreme, S.70 grows laterally immediately, forming an aggregate made of only a few particles having the shape of a crumpled distorted layer. The intermediates as well as the heterogeneities (Fig. 26) are attributable to changes or instabilities of growth inside the furnace.]

It is evident that microtexture is at the origin of the peculiar aggregation of each carbon black into single spheres, chains, or three-dimensional arrangement. Correspondingly, the physical properties of the materials depend on this peculiar aggregation. It is convenient for qualifying quantitatively the aggregation to consider the fractality [57,58]. All samples of conductive blacks have mass fractals D_M and surface fractals D_S anticorrelated. The smaller D_M,

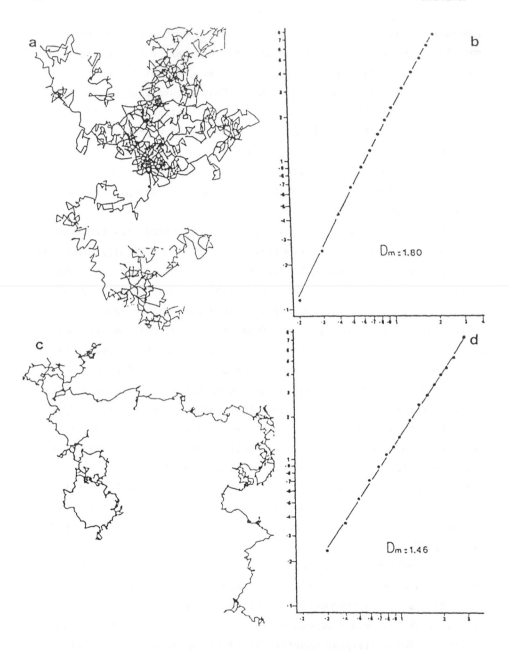

FIG. 28 Determination of mass fractals D_M for conductive carbon
blacks and model of Brownian motion: (a) and (b) poor-quality sample
and (c) and (d) high-quality sample (S.70). (From Ref. 5.)

the richer the sample in type 5 (lateral growth) and the better the material (Fig. 26). The minimum value of D_M for lateral growth is explained by the Brownian persistent flight of Mandelbrot [58]. If the successive movements are random, the fractal dimension D_M is 2 [Fig. 28(a)]. If each of them tends to depend on the preceding one, the fractal dimension tends to 1 [Fig. 28(c)]. For Ensagri (D_M = 1.8) the corresponding Brownian movement is sketched on the right of Fig. 28(a). For S.70, D_M is 1.46, corresponding to a more linear Brownian movement [Fig. 28(d)]. In conclusion, the most compact aggregate (D_M large) is also the smoothest (D_S small) and the worst material, whereas the least compact (D_M small) and the most tortuous (D_S large) is the best. As a comparison, for a thermal carbon black (P33) D_M is 1.9-2 and D_S is 1.28: D_M is maximal, D_S minimal, the specific surface is minimum, and the resistivity maximum.

2. *Carbon Shells*

True or nearly spherical configurations can be obtained either in a solid phase or in "liquid" or gaseous phase. The first case is, for example, the decomposition of carbides (internal source of carbon). In the second and the third cases the source of carbon is external and the carbon deposition occurs on a noncarbon nucleus, which may or may not interact chemically with it. In the second case, the growth mechanism is similar to that of carbon blacks.

a. Carbide Decomposition. Many examples are known showing carbides being surrounded by carbon shells under decomposition. Silicon carbide, for example [59-61], may be decomposed by very strong electron irradiation or by heat treatment. Incomplete studies [60] assumed a topotactic growth of carbon with a preferred orientation of the aromatic layers upon (111) of SiC. It was then demonstrated by Comte-Trotet [61] that carbon sheets were deposited with the same thickness on any SiC crystal faces with a complete random azimuthal orientation of the carbon layers on each face. The only character-istic feature was the parallelism between (00.1) of carbon and the planar surface of each crystal face. Such a rule will be observed very commonly for other couples of carbon with other material (see

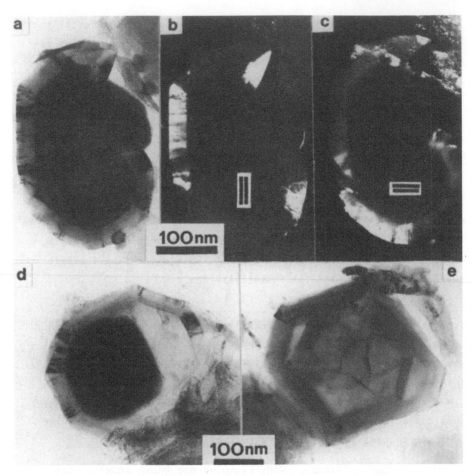

FIG. 29 Decomposition of VC. (a) BF. (b and c) Orthogonal 002 DF.
The aromatic layers are represented by a double bar. (d) Partially
decomposed VC. (e) Hollow carbon shell. (From Ref. 62.)

following sections). Another example is the decomposition of vanadium
carbide formed in heavy petroleum products by heat treatment [62]. The
initial content of V was 2-3%, present as an organometallic compound.
During heat treatment, vanadium diffuses out and small crystals of
V_2O_3 appear, trapped inside carbon pores. After further heat treatment VC is formed and then decomposed. Figure 29 shows the occurrence
of a carbon shell surrounding VC. The aromatic layers are parallel to
the VC crystal faces, as demonstrated in Fig. 29(b) and (c) (the

direction of aromatic layers is indicated by a double bar). As HTT
increases, the decomposition of VC yields at first partially hollow
and then entirely hollow carbon shells [Fig. 29(d) and (e)]. An interesting feature of these shells is the perfection of their carbon
layers. The fringes are long, stiff, and straight. The same characteristic will be found in catalytic carbons, despite the low HTT at
which they are prepared.

b. Concentric Textures Formed from Softening Phases. Ayache et al.
[63,65] heat-treated natural carbon precursors (walls of algae cells)
to 650°C at a 4°C min^{-1} rate, in sealed tubes or autoclaves, under
increasing pressures (< 2 MPa, 5 MPa, 30 MPa). Microprobe EDS analysis
shows that algae walls may contain P, Ca, and Si in variable amounts,
or even Si only. In most of the cases Si migrates alone along the
tube walls [Fig. 30(d)] in a decreasing amount near the top of the

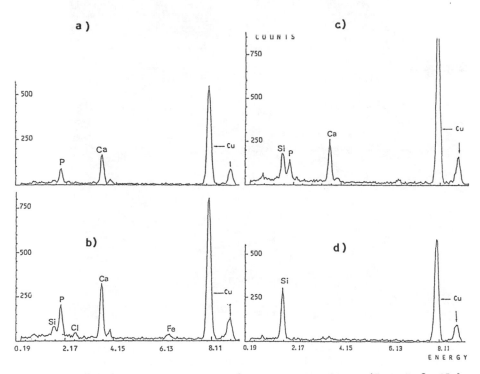

FIG. 30 Microprobe EDS analysis of algae cell walls. (From Ref. 63.)
(a) and (b) P Ca rich. (c) and (d) Si rich.

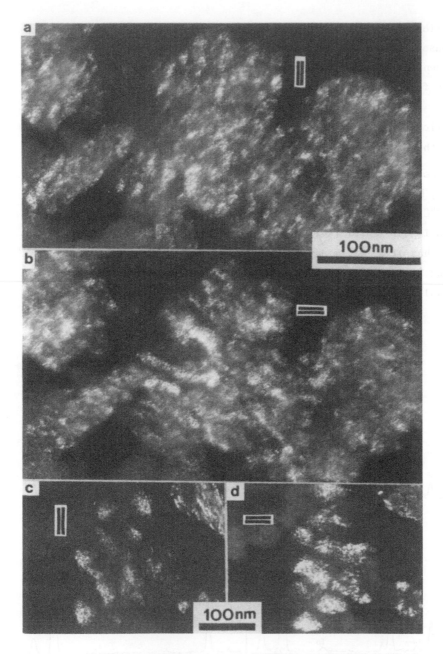

FIG. 31 Carbon black-like particles formed under pressure.
(a and b) orthogonal 002 DF and (c and d) industrial carbon
blacks. The aromatic layers are represented by a double bar.
(From Ref. 63.)

tube. P and Ca alone or mixed with the excess of Si gather in
the bottom of the tube, included in a distinct "precipitate"
[Fig. 30(a,b,c)]. For the highest pressure, the tube bottom containing a large majority of *P and Ca* yields carbon black-like particles having a concentric texture [Fig. 31(a) and (b)]. Their 002
DF images cannot be distinguished from industrial carbon blacks represented in Fig. 31(c) and (d). Such spherical particles contain themselves only P and Ca, detected by nanoprobe EDS analysis. The occurrence of carbon black-like particles prepared at a low temperature
in softening materials instead of at a high temperature (> 800°C) in
gaseous phase explains the numerous carbon black-like particles often
found in natural carbonaceous samples, such as kerogens or coals [14,
16,24,28,30,48].

At lower pressure, concentric textures appear as carbon shells
surrounding crystals. Figure 32(a) represents a BF image in which
the unknown crystals are dark (arrows). The two orthogonal 002 DF
[Fig. 32(b) and (c)] show the concentric texture also demonstrated
by 002 LF (Fig. 33, single arrows). Very often the crystals themselves yield fringes (Fig. 33). Every crystal (analyzed by EDS with
nanosonde) contains P and Ca or a very small amount of Si in addition.
The SAD patterns corresponding to areas similar to that of Fig. 32(a)
cannot be used to identify the crystals because their reflections are
not numerous enough to give complete Debye-Scherrer rings. However,
all of them are compatible with calcium phosphates and apatites.
Among all possible reflections of this family, large d spacings are
found from 1 nm down to about 0.8, 0.6, and 0.5 nm. In Fig. 33(c)
two crystals show fringes, the optical diffraction pattern (ODP) of
which is represented in the inset. The larger crystal shows a period
of 0.812 nm, the smaller crystal 0.346 nm. Both families of lattice
planes being parallel, the 0.812- and 0.346-nm periods are along the
same scattering vector in the ODF. These crystals contain only P and
Ca. They are thus phosphates.

In such materials no catalytic effect, strictly speaking, can be
assumed, since the crystals only serve as substrates for carbon BSU
deposition and no reaction repeats itself as in catalysis. Neverthe-

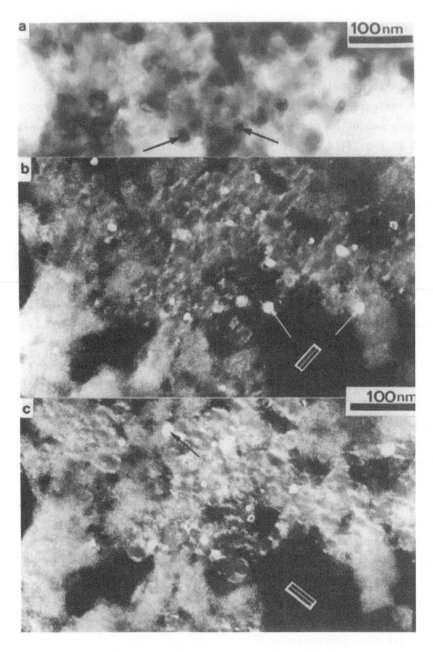

FIG. 32 Concentric shells of carbon around crystals (arrows).
(a) BF and (b and c) orthogonal 002 DF; the aromatic layers are
represented by a double bar. (From Ref. 63.)

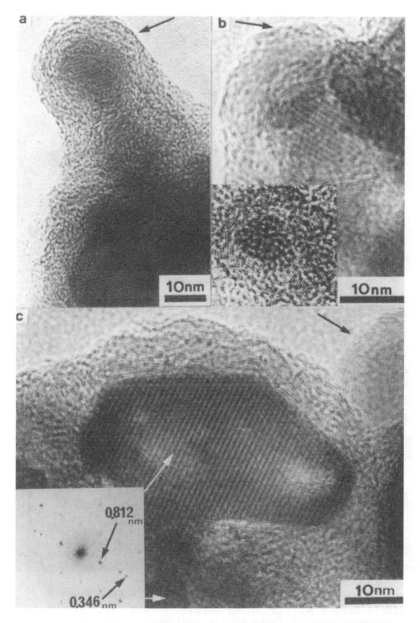

FIG. 33 Concentric shells of carbon (single arrows) around calcium phosphate crystals, 002 LF. (a) and (b) Crystals with fringes. (c) Two crystals give fringes of 0.8 nm spacing (large crystal) and 0.346 nm (small crystal) measured by ODP (inset). (From Refs. 63 and 65.)

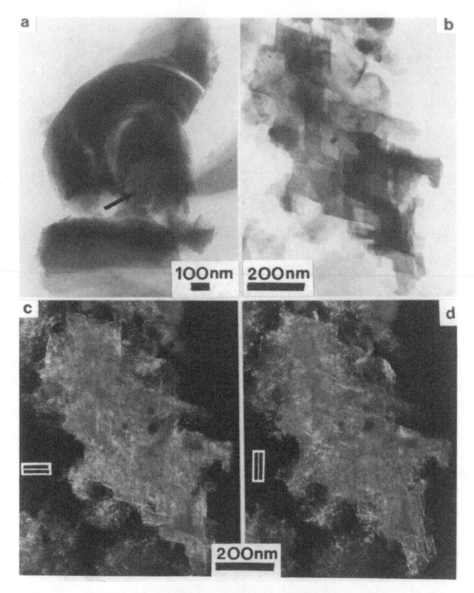

FIG. 34   (a) Concentric deposit of carbon around a carbon black
nucleus (single arrow) BF.   (b, c, d) Natural carbon replica of a
destroyed carbonate crystal (precambrian sediment), with (b) BF and
(c and d) orthogonal 002 DF.   The aromatic layers are represented
by a double bar.   (From Ref. 66.)

less, the carbon layers are relatively stiff and perfect, which
implies a high degree of organization for a relatively low temperature (650°C). Thus the crystal nucleus is not entirely inactive
with respect to carbon nucleation. We can propose for this peculiar
interaction between a pitch and a crystal the term "modification."
The crystal is the *modifier* helping the early formation of a well-
organized but not graphitized solid, in the form of a carbon shell.

Numerous impurities may act in a similar manner when present
inside a pitch heat-treated in a confined atmosphere (p > 0.1 MPa).
As an example, clays such as chlorites induce in a pitch heat-treated
at 322°C in a sealed tube a relatively thick deposit of carbon layers
parallel to the (001) faces of the chlorite crystals.

Carbon spheres themselves can be efficient as a substrate,
either inside a soaked pitch or in gaseous phase [Fig. 34(a)], whether
the nucleus itself has a concentric or radial texture.

The ability of many minerals to be embedded in a shell of carbon
may explain the occurrence of many crystal replicas found in natural
carbonaceous samples. Pyrite (FeS_2), rutile (TiO_2) [14], and also
many carbonates [66], destroyed or still present in the sediment, are
evidenced by their carbon replica. Figure 34(b,c,d) respectively
shows the BF and two orthogonal 002 DF of a carbon replica of an Mn
or MgCa carbonate. The material is a precambrian bitumen from the
region of Oklo (Gabon). The BSU of the carbon shell are not very
well oriented parallel to the crystal faces. The bright rims along
the crystal contours are thus not really complementary [Fig. 34(c,d).

c. Catalytic Shells Formed in Gaseous Phase. This is the domain of
"catalytic" carbons. It will be seen in Section IV that most of these
carbons are filamentous. They are obtained by heat treatment of a
gaseous phase (CO or CH_4, for example) in contact with catalytic particles such as iron, nickel, cobalt, and their alloys. The literature
on that subject is very abundant, and we do not intend here to make
an exhaustive review (see references in Section IV).

We will only focus on concentric textures around a metallic
nucleus [67,68]. Depending on the temperature and concentration of

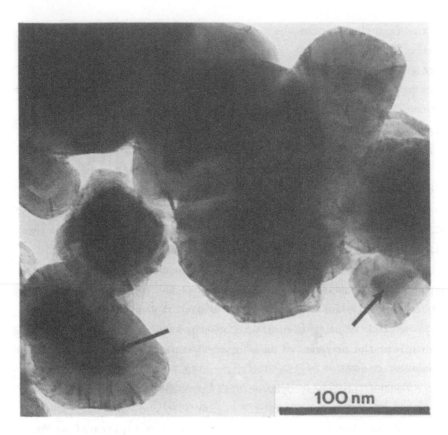

FIG. 35 Catalytic carbon shell (disproportionation of CO), HTT
650°C. Fe-Ni alloy (arrows) BF. (From Ref. 67.)

gases, metallic catalysts may get surrounded by carbon shells instead
of giving rise to filaments. The images thus obtained are absolutely
similar to that observed in Section IIIA2a (carbide decomposition),
except that the catalyst crystal always occupies the whole internal
space available (Fig. 35). As a matter of fact, it is stable during
shell formation and catches carbon from an external source instead of
decomposing itself. As in the case of carbide decomposition shells,
the carbon layers are perfect [Fig. 36(a) and (b)]. However, the
carbon is only partially graphitized. Under heat treatment graphitization does not progress, i.e., this carbon is thermally stable up
to 3000°C [67-69].

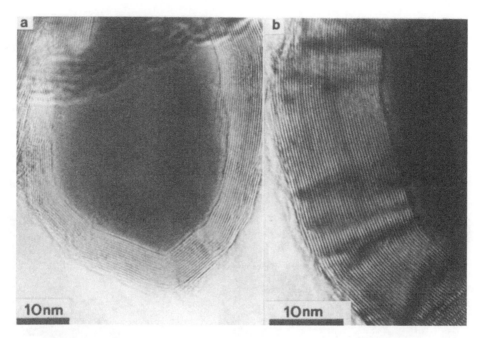

FIG. 36 Same specimen as in Fig. 35, 002 LF. (From Ref. 67.)

B. Radial Spherical Texture[*]

Pure radial microtexture was first described by Hishiyama et al. [2].
They produced spheres of this new type of materials by carbonization
under pressure (30 MPa) of mixtures of polyethylene and polyvinyl
chloride [70-72]. These authors entirely predicted the 002 DF images
due to this peculiar texture already sketched in Fig. 1(b). It was
the first case of isolated spheres proved not to have the well-known
Brooks and Taylor texture (see references in Section IIIC1b on lamellar
textures). The authors did not discuss the possible causes of such a
drastic change. Was it due to pressure, to a highly aliphatic precursor, or to another cause? Unfortunately, carbonization under
pressure of aromatic compounds [73,74] yields spheres, the texture

---

[*]It must be mentioned that despite the high temperature (650°C) at
which some of these were obtained, the materials here described are
plastic and *all the textures* are destroyed if grinding instead of
thin sectioning is used to prepare the specimens for TEM.

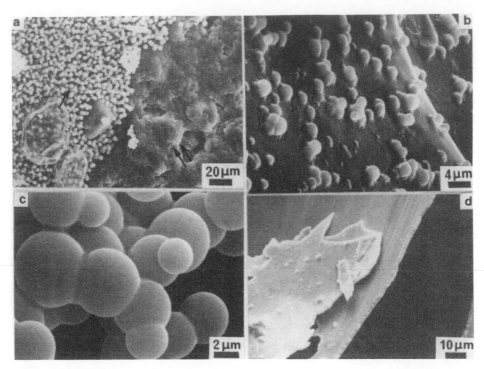

FIG. 37 Pyrolysis of algae cell walls (HTT 650°C, 30 MPa). Opened
gold tube. Direct observation by SEM. (a) Bottom of the tube, (b
and c) middle part, and (d) upper part. (From Ref. 65.)

of which was not clearly determined. It is the work of Ayache [64,65]
that brings a partial answer to that question. The precursor used was
aliphatic, the experiments were carried out under pressure (2-30 MPa),
and various impurities were present in the bulk.

In a preceding section, this work was used to illustrate an example of concentric texture favored by impurities (P, Ca). However, it
was mentioned that the residue could also contain silicon, either
alone [Fig. 30(d)] or associated with P and Ca in a larger amount
than P and Ca [Fig. 30(c)]. In the case of deposits containing only
Si, no concentric texture occurs but only radial ones. The micrographs of Fig. 37 show the inside of the gold tube serving as a reactor at 30 MPa after its opening. The SEM observations were made

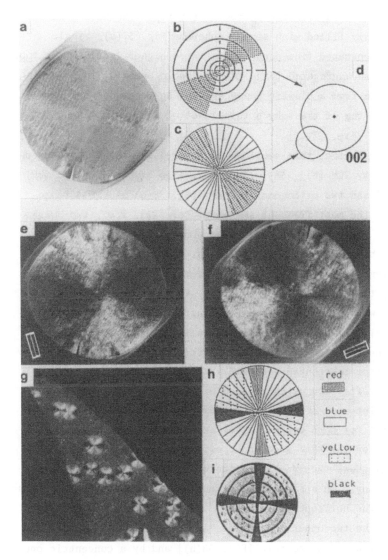

FIG. 38 Thin section of a single sphere of Fig. 37(a). (a) BF,
(b and c) sketches of concentric and radial texture, (d) aperture
location on 002 ring, (e and f) orthogonal 002 DF, (g) OM of the
embedded block (residue of thin sectioning), with the vertical
branch red (magenta), the horizontal one black, NW and SE quadrants
yellow, and others blue, (h) sketch of a radial texture (from Ref.
64), and (i) sketch of a concentric texture (from Ref. 92).

directly on the tube from the bottom to the top. The bottom of the
tube is entirely filled with single spheres [Fig. 37(a), single
arrow] well separated from each other. At a higher level in the tube
the spheres are aggregated [Fig. 37(a), double arrow]. In some places
spheres emerge from a plastic phase and are aggregated [Fig. 37(b),
(c)]. At the top of the tube a thin layer of uniform thickness covers
the tube wall [Fig. 37(d)].

In a first step, thin sections of single spheres were prepared
and examined by TEM (Fig. 38). The succession of bright field (BF)
[Fig. 38(a)] and two orthogonal 002 DF [Fig. 38(e) and (f)] repro-
duces one of the most frequent images of Ref. [2], i.e., two bright
sectors. At first sight no difference occurs with 002 DF of concentric carbon blacks except that the bright areas do not occupy the
same positions [Fig. 38(b,c,d)]. As Figure 38(c) fits well with
Fig. 38(e), the texture is radial. All other figures corresponding
to Ref. [2] were also found [see, for instance, arrow in Fig. 39(b)].

The radial texture was confirmed on larger spheres by OM
studies. The embedded block of the material that remains in the
microtome after sectioning was used as a polished section and was
observed by reflection using crossed Nicols and a first-order λ
plate. The interest of this new preparation technique is to obtain
the same area for both optical observation (by reflection or by
transmission on the thin section itself) and TEM observation.
Figure 38(g) shows the OM observation of an embedded block used for
thin sectioning. In a single sphere the vertical branch of the cross
is magenta, the horizontal one black. The NW and SE quadrants are
yellow, and the two other ones blue. If this figure is compared to
that given by a radial texture [Fig. 38(h)] and by a concentric one
[Fig. 38(i)], the radial texture is the only one that fits well.

In a second step a thin section was cut normal to the bottom
of the tube, i.e., normal to the plane of Fig. 37(a). It shows
(Fig. 39), from the bottom to the top, coalesced spheres and then
single ones. A continuous layer, in contact with the tube wall

appears underneath the aggregated spheres in the bottom of the tube.

A successive series of thin sections realized from the middle to the top of the tube was then examined in BF [Fig. 40(a)] and two orthogonal DF [Fig. 40(b) and (c)]. This figure shows first that single spheres are still visible without losing their radial texture, whereas at the extreme top of the tube (upper part of Fig. 40) a thin layer of uniform thickness is produced in which aromatic layers are at first normal to the gold tube wall. The same figure also shows a new feature. The wall of the tube is covered by a thin layer of uniform thickness, where the aromatic layers are oriented parallel to the tube wall. The change of orientation from radial to parallel by approaching the tube wall illustrates very well the tendency of BSU to lie flat on any substrate they "see" (see Section IIIA2a). Direct EDS analysis of the tube itself by SEM shows the Si content to be zero at the top and to increase down to the bottom. This increasing purification is probably simultaneous with a progressive change in the composition of the residue. Elemental analysis showed an H/C atomic ratio higher at the bottom of the tube than at the top.

Samples issued from precursors where Si content is very high were also examined by SEM. The bottom of the tube [Fig. 41(a)] contains large spheres (single arrow) and a foam of smaller spheres (double arrow). TEM data are represented in Fig. 41(b) and (c). All the large spheres (single arrow) have a radial texture, and nanoprobe EDS analysis indicates the presence of a small amount of silicon. The foam is made up of a mixture of small radial spheres, either single or associated, and *carbon shells having mainly a radial texture* (double arrow). Figure 42(a,b,c) represents a thin section of a single carbon shell, the center of which is occupied by a hexagonal crystal of SiO_2 [see also Fig. 41(b)]; (b) and (c) are orthogonal DF showing the radial texture. The crystal was identified by nanoprobe EDS analysis (only Si and O present).

The result of these experiments suggests that *the presence of Si as an impurity* does not inhibit the establishment of radial

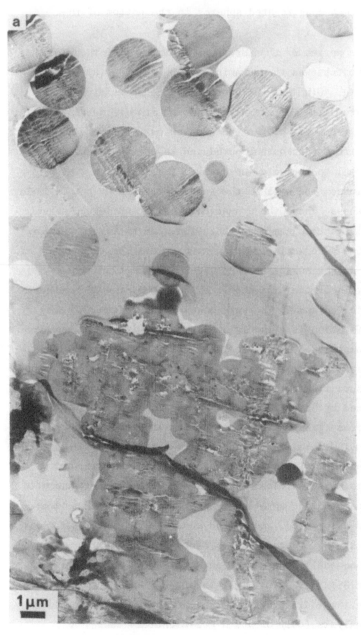

FIG. 39 Thin section of the gold tube bottom [see Fig. 37(a)].
(a) BF and (b) 002 DF. The aromatic layers are represented by a
double bar. (From Refs. 63 and 65.)

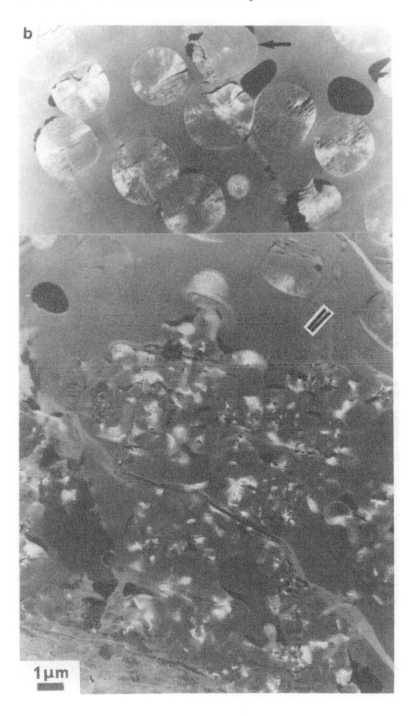

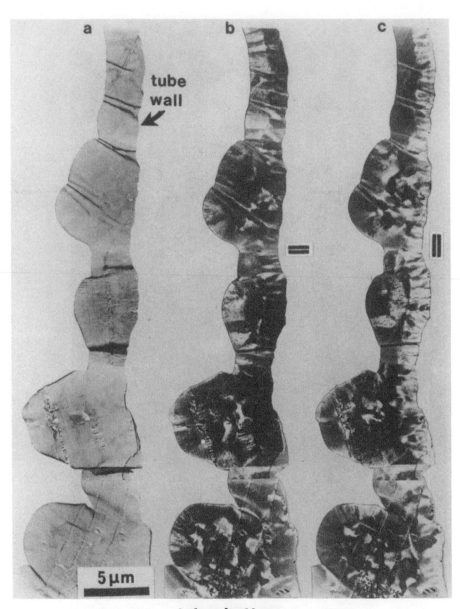

tube bottom

FIG. 40 Thin section from the middle to the upper part of the gold tube. (a) BF and (b and c) orthogonal 002 DF. (From Refs. 64 and 65.)

FIG. 41 Precursor very rich in silicon. Opened gold tube.
(a) Direct observation by SEM of the tube bottom (single arrow,
large spheres; double arrow, foam). (b) Thin section of the tube
bottom, BF. (c) 002 DF of the same area.

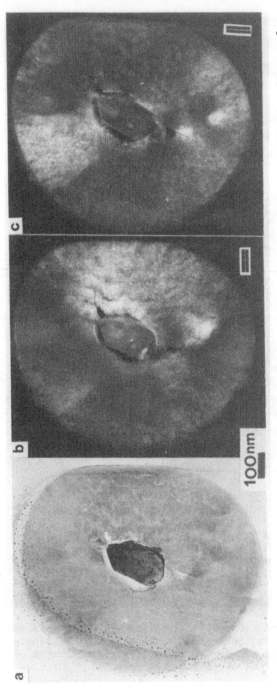

FIG. 42 (a,b,c) Thin section of a single shell: (a) BF and (b and c) orthogonal 002 DF. (From Refs. 64 and 65.)

texture, whereas P and Ca fixed inside a shell favors or even creates the concentric texture. As the impurity content is large enough for obtaining phosphate or silicophosphate crystals, the carbon shell formed is still concentric, whereas for silica crystals the shell is mainly radial. As the aromatic layers of a radial shell approach the silica crystal they are deflected and end as a thin layer parallel to silica. New experiments are being conducted for studying the texture of pure aromatic and aliphatic compounds heat-treated under pressure. They show the utmost influence of aromaticity [64,65].

C. Statistically Spherical Textures

A large number of carbonaceous materials show this texture. Among them are the electrically conductive carbon blacks already described in Section IIIA2b, most of the carbons obtained from softening phases, some pyrocarbons, and some carbons heat-treated under pressure. As already shown in Section IIIA2b, such texture implies a model of continuous crumpled sheets of carbon layers having an average radius of curvature r. There is no reason for r to be either constant or small. When it tends to infinity, lamellar textures are obtained. It is thus impossible to separate these peculiar types from the others, so that porous and lamellar carbons will be considered together in this section.

1. *Materials Issued from Softening Phases (Mainly Aliphatic)*
a. *Introduction.* Most of the organic materials (composed of C, H, O, S) when pyrolyzed under an inert gas flow follow a similar trend [3, 12,13,75]. First, during the primary carbonization they lose oxygenated functional groups as well as their sulfur and nitrogen equivalents (CO_2, H_2O, H_2S,...). Then they lose aliphatic CH groups, forming tars. At the end of tar release the material is a brittle solid, whereas during the first steps it was plastic or viscoelastic. As HTT is increased (secondary carbonization), gas release and cokefaction occur.

The singular point separating the primary from the secondary carbonization can be determined by considering chemical composition and rheological property variations versus HTT. IR analysis, for example [76], shows that the end of CH aliphatic groups release (tar release) corresponds to an extremum in the aromatic CH groups, which then release also (gas release) (Fig. 43). Figure 43 shows the comparison between aliphatic and aromatic group behavior [76] during pyrolysis of purified *Lycopodium* spores $C_{90}H_{144}O_{27}$ [77]. If Fig. 43 is compared to ESR data [14,29], the extremum in aromatic CH immediately follows the maximum of spin concentration. In addition, Fig. 43 also shows an increase in the temperature of LMO occurrence as the pressure increases. This will be explained in Section IIIClc.

The work of Fitzer et al. [78-80] gathers a series of interesting rheological characteristics of more or less heavy carbonaceous products (all of them are hydrogen-rich, i.e., they are light products). These authors studied different fractions of a pitch and various pitches for which they measured apparent viscosities, carbon yields, and quinoline-insoluble percentage versus HTT. Figure 44 shows part of their data. The curves marked (toluene-soluble) correspond to the lightest products. From this figure, the point corresponding to the brittle solid state is

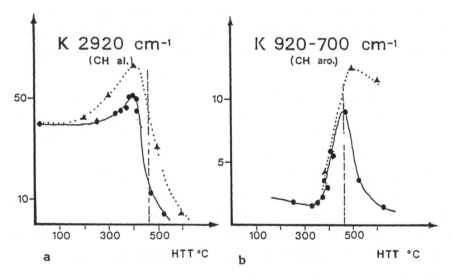

FIG. 43 IR spectroscopic analysis of sporollenin pyrolysis (standard conditions, full lines; under 0.5 MPa, dashed lines). (a) Aliphatic CH groups. (b) Aromatic CH groups (vertical line corresponds to LMO occurrence in standard conditions). (From Refs. 29 and 76.)

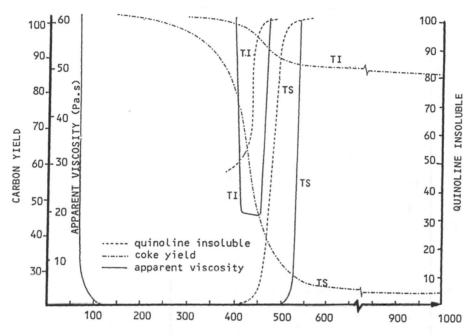

FIG. 44 Variation of some physicochemical features during pyrolysis of fractionated pitches. Apparent viscosities (solid lines), carbon yields (dashed lines), quinoline insolubles (dotted lines). TI, toluene-insoluble; TS, toluene-soluble. (From Ref. 78.)

seen to correspond to a jump of apparent viscosity to infinity. Correspondingly, the quinoline-insolubles tend to 100% and the carbon yield tends to reach a plateau. This figure also shows that for heavier materials (TI, toluene-insoluble) the softening temperature increases whereas the solidification temperature decreases. The carbon yield increases. The quinoline-insolubles appear at a lower temperature and also reach 100% at a lower temperature [81]. This tendency is more marked for the heaviest materials (oxygen-rich materials) and also for some coals [12,13]. Most of them, though visco-elastic, do not soften appreciably, so that only microhardness can be measured (Fig. 45). They have a very high carbon yield and are nearly entirely insoluble in quinoline. However, when considered according to their increasing initial carbon content, their micro-hardness also shows a jump toward infinity for the highest ranks.

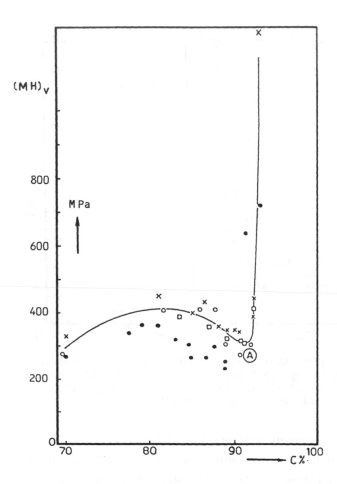

FIG. 45 Vicker's microhardness index of coals. (From Ref. 12.)

The end of carbonization, the pure carbon stage, is not well marked, since the estimation of the purity of carbon depends on the way elemental analysis is performed. In fact, accurate elemental analysis shows that heteroatoms such as H, O, S, and N can remain firmly fixed in carbons (Fig. 46) up to 2000°C and sometimes beyond [82]. It will be seen later that TEM may help to better define this point.

When HTT is increased, carbon layers rearrange themselves and eventually graphite may be obtained depending on the elemental composition of the precursor.

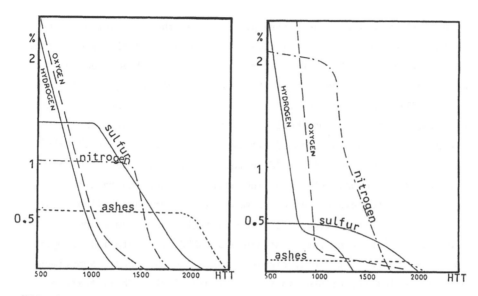

FIG. 46 Elemental analyses of two petroleum cokes. (From Ref. 82.)

b. Hydrogen-Rich Materials. i. Carbonization. Among others,
hydrogen-rich materials comprise light pitches (coal tar pitches or
petroleum pitches) and light petroleum products issued from light
oils. The optical studies of their carbonization have yielded an
abundant literature [12,13,81,83-88]. Results can be summarized as
follows. Inside the optically isotropic pitch, simultaneously with
the occurrence of quinoline insolubles, optically anisotropic spheres
appear (mesophase spheres). They grow, then coalesce. As the pitch
entirely disappears, the coalesced spheres solidify into bulk mesophase and form coarse mosaics. The latter are formed of optically
isochromatic distorted areas, the size of which may reach 200 µm for
the lightest materials. Because of the poor resolution of OM (at
best a fraction of a micrometer), optically isotropic materials cannot
be considered as free from mesophase. Moreover, the nucleation of
mesophase spheres cannot be observed.

Ihnatowicz et al. were the first ones to observe single spheres
of mesophase with TEM in DF [81], whereas Brooks and Taylor performed
SAD [84]. More recently, Auguié et al. [18,89] described in detail
the texture and microtexture of single spheres and mosaics. However

none of them followed the entire process from the nucleation to the
formation of cokes. This was studied by Bonnamy [90,91] by comparing
OM and TEM data. The latter references [90,91] will be used to
describe in detail the carbonization behavior of hydrogen-rich materials. The example chosen is an atmospheric residue (residue of the
oil distillation at atmospheric pressure) of a light Arabian oil
(ALAR). ALAR was heat-treated either at $0.5°C \ min^{-1}$ or at $4°C \ min^{-1}$
up to 1000°C by steps of 5-10°C in the vicinity of single-sphere
occurrence. The only effect of increasing the heating rate is to
increase the temperature at which mesophase occurs and collapse the
various interesting steps [81]. The data given here correspond to
a $0.5°C \ min^{-1}$ heating rate. At 430 and 435°C, the material is optically isotropic. At 440°C, single spheres > 6 μm appear, sometimes
already coalesced, diluted in the isotropic pitch. Their texture
verifies the Brooks and Taylor model (Fig. 47). Figure 47(a) shows
the succession of optical images obtained on a polished section as
the specimen stage is rotated [the same reasoning is applied as in
Fig. 38(h) and (i) [92]]. When two single spheres coalesce, the image
becomes more complicated because disclinations [93-96] are introduced
at the boundaries [Fig. 47(b)]. Up to 450°C, the coalescence pro-
gresses. The mosaics formed later continue to grow above 485°C.
After 500°C, coarse mosaics are produced that do not grow further.
They were observed at 1000°C.

TEM studies were performed on thin sections. At 430°C, the
material, though optically isotropic, shows ellipsoidal anisotropic
nuclei [Fig. 48(a) and (b)] diluted in an isotropic matrix containing
BSU at random. In 002 DF all these nuclei appear bright or dark
depending on the location of the objective aperture. The same nucleus
is either entirely bright [Fig. 49(a)] or entirely dark [Fig. 49(b)]
for two orthogonal positions of the aperture. The aromatic layers
are therefore all parallel to each other and normal to the external
contour of the nucleus. This would not be expected if they followed
the Brooks and Taylor model. The smallest detectable nucleus was
about 30 nm in size and the contrast of the bright image was very

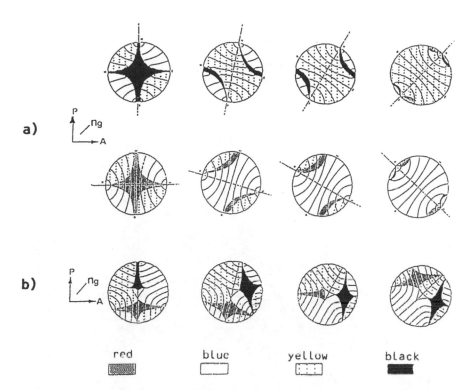

a)

b)

red   blue   yellow   black

FIG. 47 OM of mesophases (crossed polarizers, first-order λ plate
added). (a) Single sphere during stage rotation. (b) Coalesced
sphere. (From Ref. 92.)

poor [Fig. 48(a) and (b)]. As HTT increases, the density, the size
and the contrast of nucleus increase [Fig. 48(c)]. In this figure,
the fact that nuclei are elongated at random implies that their shape
is not due to stresses developed in the thin section. As early as
440°C, single spheres of various sizes appear. All of them verify
the theoretical 002 DF images predicted by Auguié et al. [89].
Figure 50(a,b,c) represents a succession of the most frequent 002 DF
images above the predicted image sketches. Figure 50(e,f,g) represents in the same manner the succession of images obtained from Fig.
50(a) by displacing the aperture counterclockwise along the 002 ring.*

---

*Note that the 002 DF images of a meridian section where all the layers
fulfill the 002 Bragg condition is equivalent to the corresponding
polished section viewed by OM. Figure 50(a,e,f,g) is thus comparable
to the lower line of Fig. 47(a).

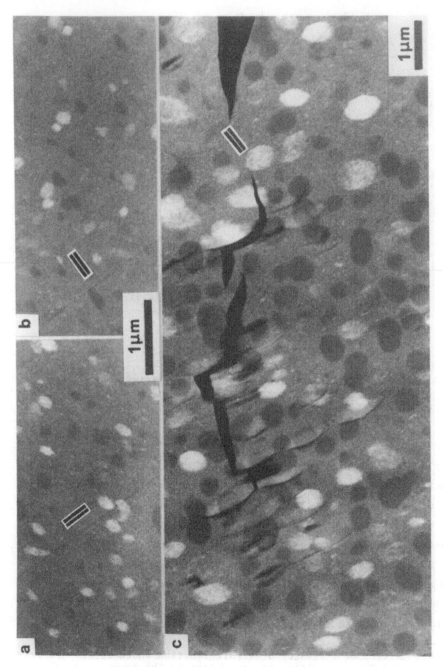

FIG. 48 Mesophase nuclei. Thin section. (a and b) Orthogonal 002 DF (low density of nuclei).
(c) 002 DF (high density of nuclei). The aromatic layers are represented by a double bar.
(From Ref. 90.)

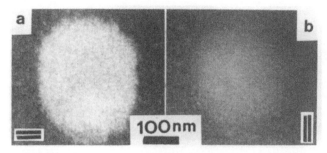

FIG. 49 (a) and (b) Nucleus imaging in orthogonal 002 DF. The
aromatic layers are represented by a double bar. (From Ref. 90.)

The single spheres obey the Brooks and Taylor model. From an ellipsoidal nucleus to a single sphere, the aromatic layers have to curve
in order to remain normal to the external surface, though being
approximately parallel to each other. It seems that the unique rule
for the whole growth is that aromatic layers "avoid" being wetted on
their faces by the pitch and ensure lateral growth. Correspondingly,
the nuclei have to be ellipsoidal when they are small and the layers
have to curve during growth, insuring acquisition of a spherical
shape. Such a mechanism recalls the experiments of Diefendorf on
the wettability of HOPG by pitch. Pitch wets the edges of the layers
and remains as a drop on the face. Figure 51 illustrates the simultaneous occurrence of nuclei, single spheres, and coalesced spheres
(arrow) in the same sample.

Optical studies on liquid crystals [97-101] described a columnar
order of piled up disk-like molecules (discotic order) superior to the
order observed in nematic mesomorphs. However, the optical studies of
mesophase [102,103] conclude to a nematic order in a sphere. The
arrangement of BSU inside mesophase spheres was studied by Auguié et
al. [89], first in noncoalesced single spheres, then on coalesced
ones. Figure 52(a) shows one part of a single noncoalesced sphere;
this was verified by the fact that, observed by DF, the sphere gives
one of the images of Fig. 50. The parallel set of fringes associate
face to face, forming distorted columns of 8-15 fringes, similar to

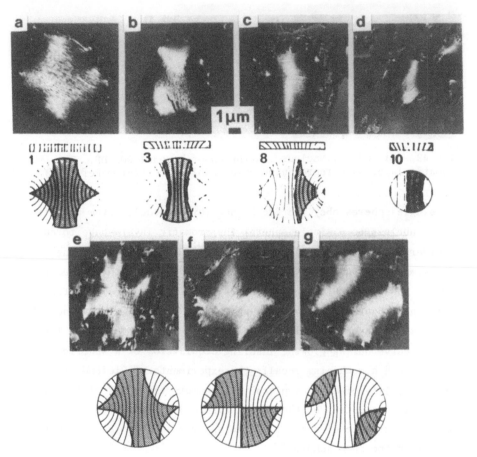

FIG. 50 Most frequent 002 DF images of mesophase noncoalesced spheres.
(a,b,c,d) Sketched in 1,3,8,10 (in the upper part profile of the corresponding perpendicular slice). The dashed areas limit the area bright
in 002 DF. (a) Meridian section, (b) section parallel to and above (a),
(c) section inclined 15° relative to the NS pole line and cut at ordinate 0.4, (d) same section cut at ordinate 0.8 (near the vertex of the
sphere), and (e,f,g) image of (a) as the aperture is displaced counterclockwise along the 002 ring [compare to the lower line of Fig. 47(a)].
(From Ref. 89.)

thick piles of plates. Despite their distortion the columns yield

highly oriented ODP [Fig. 52(b)]. It is thus typically a columnar

order, i.e., a discotic order. This order is very fragile and sensi-

tive to stresses, since it is immediately destroyed by coalescence

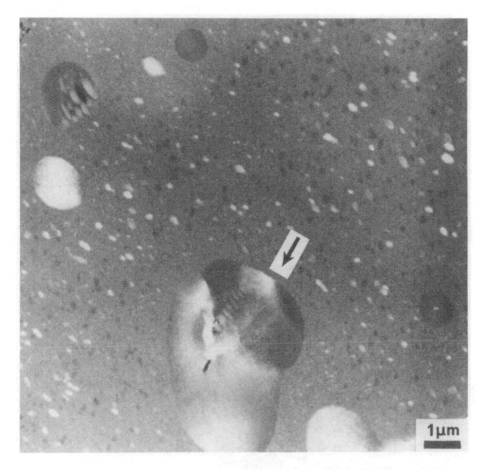

FIG. 51 Thin section of mesophase growth in ALAR. 002 DF; nuclei,
single spheres and coalesced spheres (arrow). (From Ref. 90.)

between two spheres and decreases to nematic order. Figure 52(c)
corresponds to a coalesced sphere of mesophase. Only single BSU
(two to three fringes) are associated edge to edge and face to face
with tilt and twist boundaries. The ODP [Fig. 52(d)] shows an increasing misorientation corresponding to the opening of the reflections into arcs. Correspondingly, x-ray studies [104] show a decrease
in order from single spheres to coalesced ones. In the first case,
the stacks are 8-15 layers thick and their 002 spacing is 0.34 nm.

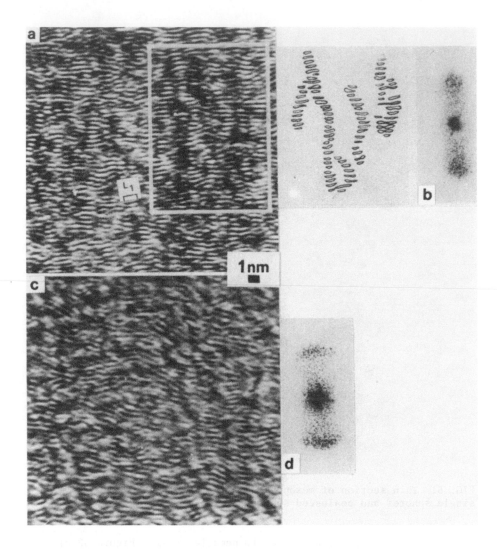

FIG. 52 Thin section of mesophase, 002 LF. (a) Single uncoalesced
sphere. (b) Sketch of the zone of the ODP and ODP of (a).
(c) Coalesced sphere. (d) ODP of (c). (From Ref. 89.)

In the second case, the stacks comprise only two to three layers,
with a larger 002 spacing (0.344 nm). The apparent discrepancy with
OM studies is explained by the manner in which mesophase growth is
observed. The OM in situ observation is performed by means of a
heated stage where the mesophase nucleation occurs inside a thin

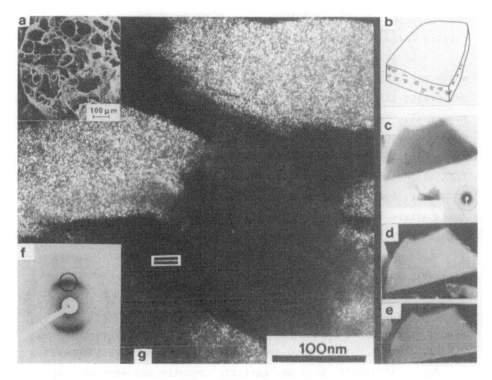

FIG. 53 (a) Porous semicoke. (b) Fragment of a pore wall (sketch).
(c) Lamella lying flat. BF (inset, SAD pattern). (d and e) Orthogonal 002 DF. (f) SAD pattern of a lamella oriented edge-on (the
aperture location is circled). (g) 002 DF of (f). The aromatic
layers are represented by a double bar.

film. Therefore it coalesces at the very moment when becoming visible.
The discotic order is thus missed. The decrease in order due to coalescence is well explained by the stresses introduced, which are materi-
alized by disclinations [93-96].

   At the moment when coarse mosaic forms, the material acquires
very large domains where the aromatic layers are oriented approximately
in parallel [90,91]. Each domain corresponds to an optically isochromatic area, the orientation of which progressively changes in the
compact material [see Fig. 53(g)]. Their shape is roughly isometric.
First just before the solidification, a macroporosity occurs. Then,
after solidification, the oriented domains become thinner and form
lamellar pore walls [Fig. 53(a)], inside which the aromatic layers

are approximately parallel [Fig. 53(b)]. Because the material is a
brittle solid it is not absolutely necessary to prepare the specimen
for TEM by thin sectioning: simple grinding is suitable. The pore
walls are thus broken into lamellae, which tend to lie flat on the
supporting film [Fig. 53(c)]. Because the layers are parallel to the
pore wall, they are also parallel to the lamella plane. In 002 DF
such particles are never bright [Fig. 53(d) and (e)]. On the contrary, if they are seen edge on [Fig. 53(g)], the aromatic layers
fulfill the 002 Bragg condition and the SAD pattern contains two 002
arcs [Fig. 53(f)]. The lamella becomes full of bright dots (BSU) as
the aperture is centered on the arc [Fig. 53(g) (upper figure)]. It
becomes dark for the orthogonal position of the aperture (lower figure).

   ii. Graphitization. All materials practically free of oxygen
or stable sulfur are graphitizing carbons. The progressive thermal
graphitization was first followed by TEM by Oberlin and Terrière [22]
on anthracene-based carbons (AC). Then, it was followed in more
detail, also on AC, by Rouzaud [16,23]. Among graphitizing carbons
another good example is thin or relatively thin carbon films [31,32,
105,106]. At 1000°C, they are entirely aromatic and very well oriented
with their carbon layer stacks (BSU) parallel to the film plane (see
Fig. 11). A single LMO area occupies the whole film. Except for
minor differences in the temperature range of the various graphitization steps, all graphitizing carbons follow the same trend and obey
the same mechanism.

   After primary carbonization, which yields large LMO of nearly
parallel BSU, the distortions in the sheets of layers eliminate by
steps. Each of them corresponds to the release of a given type of
defect.

   Careful measurements were made both on AC and carbon films (CF)
to obtain the type of data mentioned in Table 3. Figure 54(a) and
(b) correspond to the thickness data of N, L'_{c004}, and l_c. Figure
54(c) corresponds to diameter data $L'_{all(0)}$, l_a, L_1, L_2, and $L_{all\ TF}$.
Figure 54(d) recalls $L_{all\ TF}$ as compared to $L_{all\ Cr}$ measured with
another scale of ordinate. Figure 54(e) corresponds to β and Fig.

TABLE 3 Possible Numerical Values Derived from TEM Data

SAD patterns	L'_{c004}	Half-width of 004 reflections
	$L'_{a11(0)}$	Half-width of 11(0) ring
002DF	L_c	Thickness of elementary bright domain (BSU)
	L_a	Length of elementary bright domain
002LF	N	Number of fringes in a stack
	L_1	Length of a perfect fringe
	L_2	Length of a distorted fringe
11DF	$L_{a11\ TF}$	Length of turbostratic moiré fringes
	$L_{a11\ Cr}$	Diameter of a domain showing rotational moiré fringes
ODP of 002LF	β	Arc opening
	ΔD_{002}	Interfringe spacing spreading

Source: From Ref. 23.

54(f) corresponds to ΔD_{002}. Each curve shows sudden changes of slope corresponding to four recognizable stages (Fig. 55) illustrated by SAD, 002 LF and 11 DF.

The *initial stage or stage 1* was already described as the beginning of secondary carbonization (see Fig. 53). The various data relative to thickness and diameter remain constant and small. Only single BSU azimuthally distributed at random are observed. However, during this stage β and ΔD_{002} decrease. The parallel ordering of BSU improves and the interlayer spacing becomes smaller and more constant. In AC carbons the *defects eliminated* during that stage are most of the *aromatic CH groups* grafted upon the BSU boundaries (at 900°C, the atomic H/C ratio is < 0.03).

Stage 2 begins with a change of slope in thickness followed by a plateau [Fig. 54(a,b)]. During that stage a partial recovery of the columnar order of mesophase single spheres occurs. Small distorted columns trapping misoriented single BSU between them [arrows in Fig. 56(a)] are observed in 002 LF and 002 DF. The diameter does

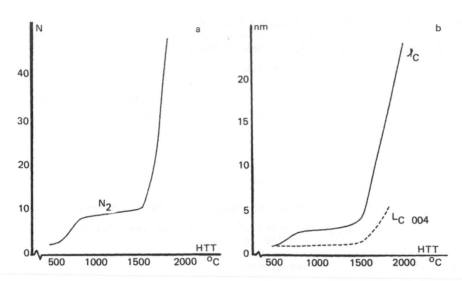

FIG. 54 Heat treatment of AC. (a) and (b) thickness data, (c)
diameter data, (d) diameter data with reduced scale, (e) misorientation β (twist), and (f) interlayer spacing spreading ΔD_{002}. (From
Ref. 23.)

not increase and remains that of a single BSU [Fig. 54(c)]. The
reason is that the layers are not coincident between two adjacent
columns. However, the formation of columns is responsible for the
sudden increase in thickness parameters. The term β, which represents the distortion of the columns, is small and stable and ΔD_{002}
decreases slowly [Fig. 54(d) and (e)]. However, the improvement of
β and ΔD_{002} is emphasized by the occurrence of 00ℓ high orders in
SAD patterns [compare Fig. 58(b) with 58(a)]. Nothing happens during
that stage except a decrease to zero of the number of single misoriented BSU. The model of discs represented in the inset of Fig.
56(a) is due to Blayden et al. [107]. It shows how the lateral
coalescence of columns is made impossible by misoriented single BSU.
This stage corresponds to an association of BSU *face to face*, i.e.,
to the *release of interlayer defects*.

Stage 3 begins after 1500°C with a very strong change of slope
of the thickness curves accompanied by a moderate increase in diameter
data [Fig. 54(c) and (d)]; β and ΔD_{002} tend to zero [Fig. 54(d) and
(e)], so the 00ℓ orders increase again and sharpen, whereas the 00ℓ

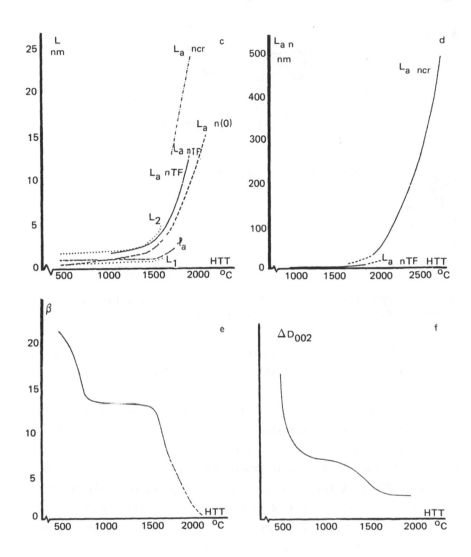

arcs close. The coalescence of the distorted columns has taken place
and longer distorted layers are produced. At the beginning of this
stage, the distortions are in some places pseudoperiodic [Fig. 56(b)],
with a period similar to the column diameter, i.e., the initial BSU
diameter (< 1 nm). Then the radius of curvature of the layers increases randomly [Fig. 57(a)]. This stage corresponds to an association *edge to edge* of the columns, i.e., to the *release of in-plane*

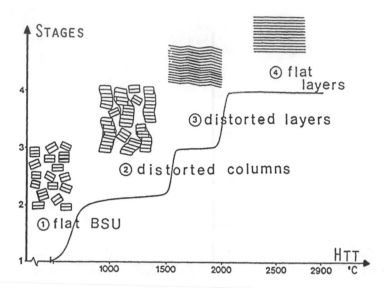

FIG. 55 Steps of increasing order versus HTT for graphitizing
carbons [stage 1 is similar to Fig. 52(c)].

defects. Up to the end of this stage the material remains turbostratic, as shown by SAD and 11 DF [see Fig. 14(a) and Section
IIB5b).

   Stage 4 begins above 2000°C after the release of all the defects.
The distortions are all annealed at the end of the stage 3. The layers
become suddenly stiff, flat, and perfect [Fig. 57(b)]. It is only
during this stage that crystal growth may begin, accompanying graphitization. The last very strong change in slope occurs only for the
diameter [Fig. 58(d)] measured in 11 DF ($L_{all\ Cr}$) [Fig. 54(d)]. All
other data cannot be obtained because the 002 Bragg condition becomes
too strict (see Section IIB1b) and the reflections in SAD patterns
too sharp, whereas L_2 tends to infinity.

   The probability P of finding a pair of layers in the graphite
order begins to increase [108-111] as measured from the modulations
of hk bands in x-ray patterns. SAD patterns also show these progressive modulations. At last hkl reflection occurs [Fig. 58(c)]. Correspondingly the turbostratic moiré fringes disappear, whereas the
diameter of hexagonal coherent scattering domains increases very

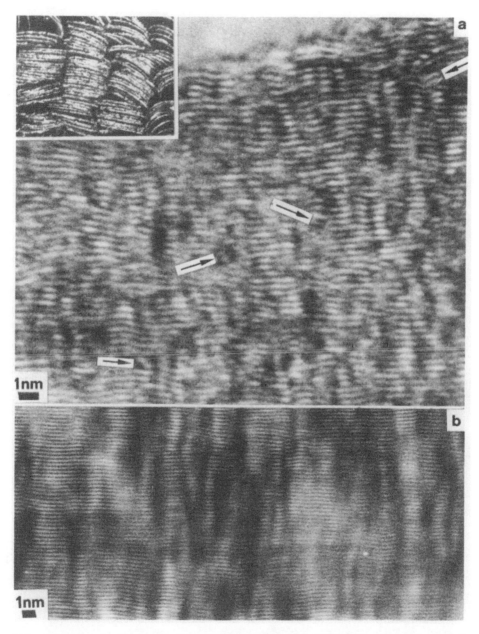

FIG. 56 Heat-treated AC, 002 LF. (a) Stage 2 (1300°C). Inset: model
of Blayden [107]. (b) Stage 3 (1800°C), small period of distortions.
(From Refs. 16 and 23.)

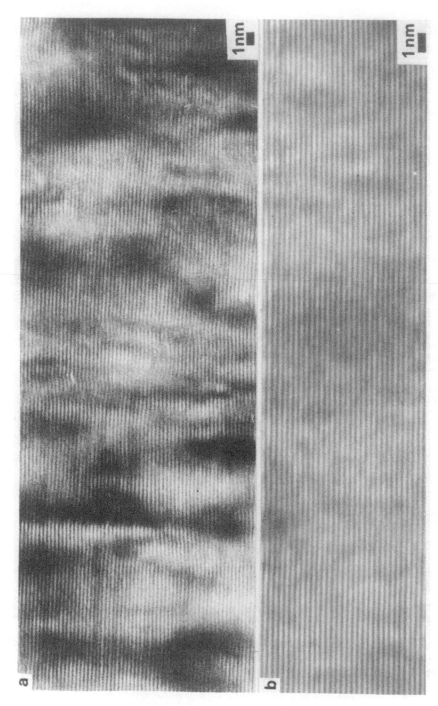

FIG. 57 Heat-treated AC, 002 LF. (a) Stage 3 (1800°C), large period of distortions. (b) Stage 4.
(From Refs. 16 and 23.)

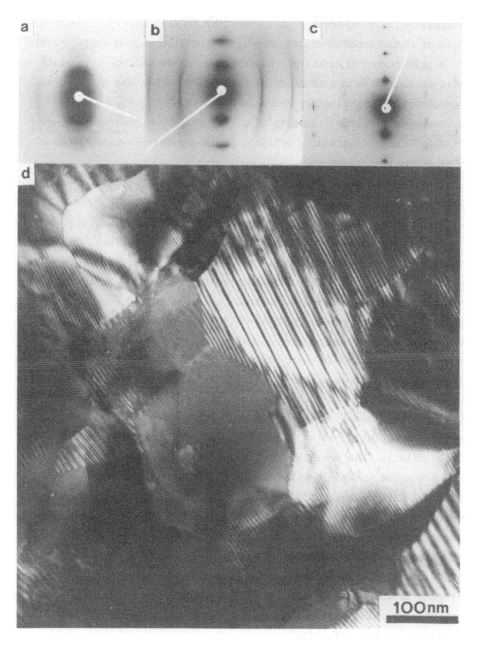

FIG. 58 (a,b,c) Heat-treated AC, SAD patterns: (a) stage 1,
(b) stage 2, and (c) stage 4. (d) 11 DF, stage 4 (2600°C).
(From Refs. 16 and 23.)

quickly [112] [compare Fig. 14(a) and (b) to Fig. 58(d)]. The thickness of the crystallites probably increases also very quickly, since in a thin carbon film the moirés disappear above 2340°C. The thickness of this film was about 10 nm. At 2340°C there are at least two superimposed crystallites, and only one at 2490°C. Figure 58(d) corresponds to thick fragments of a graphitizing carbon heat-treated at 2600°C still containing moirés. The hexagonal crystallites are over 300 nm in diameter.

Changes in microtexture are responsible for the behavior of organic matter during carbonization and graphitization. Microtexture is established as soon as the material is a brittle solid, i.e., at the end of the primary carbonization where stage 1 begins. Inevitably, when the energy added by HTT reaches the activation energy of a given defect, the modification of the layers occurs, fixing at the same time the electronic, mechanical, and structural properties of the material. The larger the LMO, the higher the ability to graphitize and the larger the refractive index, which can reach that of graphite at a very low HTT (800-900°C) for graphitizing carbons [16,23]. Correspondingly, the four stages recognized by TEM correlate with the variations of diamagnetic susceptibility, resistivity, Hall effect, and magnetoresistance, for example [113]. The average specific diamagnetic susceptibility [114] also shows four stages as HTT increases [Fig. 59(a)]. The resistivity [115] decreases drastically as the BSU get close enough to allow charge carriers to jump from one to another. It also slightly improves as all in-plane defects are wiped out, i.e., at stage 4 over 2000°C [Fig. 59(b)]. Both magnetoresistance and Hall effect [114] are sensitive only to the stage 2-stage 3 transition and more strongly to the stage 3-stage 4 transition [Fig. 59(c) and (d)]. Incidentally, the abrupt transition from stage 3 to stage 4 approximates the real end of secondary carbonization, because strictly speaking, all carbons still containing heteroatoms are cokes (see Fig. 46 and Ref. 82).

c. *Oxygen- or Sulfur-Rich Precursors* [3,14,16,17,19-21,27-30,116-121]. *i. Carbonization.* Many heavy petroleum products, saccharose,

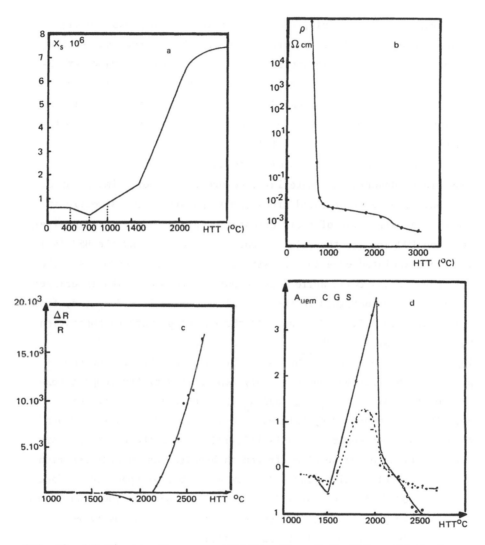

FIG. 59 (a) Diamagnetic susceptibility (from Ref. 114). (b) Resistivity (from Ref. 115). (c) Magnetoresistance (from Ref. 114). (d) Hall coefficient (full line 80 K, dashed line 300 K). (From Ref. 114.)

cellulose, many resins, and oxidized or heavy pitches generally contain oxygen, or sulfur that is stable above 1700°C (the case of labile sulfur will be discussed later). At the beginning of carbonization all of them contain BSU distributed at random. In the same manner as

in hydrogen-rich materials, the LMO nucleates, then grows [90,91].
Nucleation and growth of LMO will be studied in the same subsection
as the possibility of graphitization. At the end of primary carbonization all those materials reach the solid state. They are thus
formed of areas locally oriented (LMO), themselves statistically
distributed at random in the specimen. The DF imaging of such materials was detailed in Section IIB4.

By interpreting images similar to those of Figs. 6-10, it was
possible to describe the materials, before the end of primary carbonization, as made of oriented areas (LMO) approximately isometric in
shape. The direction of their orientation plane changes progressively
from one LMO area to the adjacent one. Inside each LMO the BSU (< 1
nm) are associated edge to edge with tilt and twist boundaries. The
tilt angle α and twist angle β are about ± 20 to $\pm 30°$. The interlayer
spacing of BSU contained in the initial precursor is widespread, from
0.8 nm down to 0.34 or 0.35 nm. At the end of primary carbonization
the spacing spreading ΔD_{002} is reduced. There is no spreading at
1000°C HTT. The model of the solid obtained just before the end of
primary carbonization can be roughly represented by the compact texture of the three LMO A_1, A_2, and A_3 [Fig. 60(a)]. The lack of
porosity is emphasized by BET measurements carried out with various
gases on a large series of coals [122-124]. Near the end of primary
carbonization, pores develop (frozen-in bubbles) as in hydrogen-rich
materials. BET studies [122-124] show a maximum of specific surface
at about 900°C, followed by a decrease to zero between 1000 and 1200°C.
However, x-ray small-angle diffusion shows that the porosity remains
but is no longer accessible to gases [125].

During secondary carbonization the rearrangement of BSU relative
to one another occurs. Inside each LMO A_1, A_2, and A_3 the tilt α and
twist β decrease (equivalent to stage 1 previously described), and
the interlayer spacing spreading ΔD_{002} decreases, so that the thickness of the LMO considerably decreases, whereas its *diameter* remains
nearly constant [Fig. 60(b)]. Because the LMO of the material are
having their orientation changing progressively from P_1 to P_2 to P_3

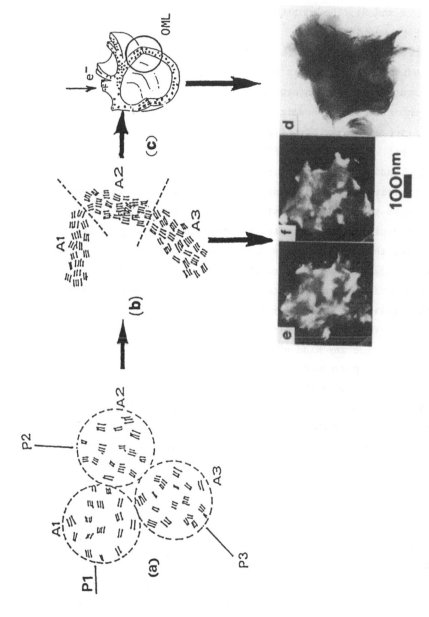

FIG. 60 (a) LMO just before the end of primary carbonization. (b) LMO after the end of primary carbonization (semicoke). (c) Formation of pores [compare with Fig. 1(c)]. (d) BF of (c). (e and f) Orthogonal 002 DF of A_1, A_3, and A_2.

(see Figs. 8 and 9), the continuity between pore walls is still
ensured [Fig. 60(b)].

The mechanism of Fig. 60(b) is repeated for all LMO. A texture
of crumpled sheets of paper is thus formed [Fig. 1(c)], rather than
a foam texture. At the contact between adjacent sheets, if two boundaries touch each other, bonding occurs and something like Fig. 60(c)
is formed (see also Fig. 6). Figure 60(d) represents a BF image
similar to Fig. 60(c), whereas Fig. 60(e) and (f) correspond to the
LMO A_1, A_3, and A_2 illuminated in orthogonal 002 DF.

It is clear that the imaging in 002 DF is easier for a coke than
for a semicoke since the contrast of the individual scattering domain
that images a single BSU increases (improvement in the parallelism of
layers). For this reason, *cokes heat-treated at 1000°C were chosen,
except in special cases, to study LMO and to evaluate their extent by
measuring the diameter of each LMO.* Moreover, since the materials
are brittle solids it is simpler to prepare the sample by grinding
than to make a thin section. It must be stressed that before the
brittle solid stage any kind of texture is entirely destroyed by
grinding, even if performed at a low temperature.

ii. Influence of elemental composition. Although oxygen- and
sulfur-rich precursors follow the same trend, they nevertheless differ
by the extent of their LMO. By using the various modes described in
Section II, not only the shape but also the extent of LMO can be
evaluated. An example of 002 DF imaging of LMO of increasing size
is illustrated in Fig. 61. Because of the irregular shape of the
clusters of bright dots observed, only a relatively wide range of
sizes can be defined for each class. The LMO can be classified relative to each other more than accurately measured. Ten classes were
thus defined (Table 4). For classes 1 to 8 the fragments observed
(obtained by grinding) are larger than each LMO. For the last two
classes, the LMO becomes larger than the fragment. The latter is
then a lamella corresponding to a fragment of mosaic. Class 10 (flat
lamellae) corresponds to the hydrogen-rich materials already studied
in Section IIIClb (Fig. 53). However, class 9 (crumpled lamellae) is
an interesting intermediate between porous materials and lamellar

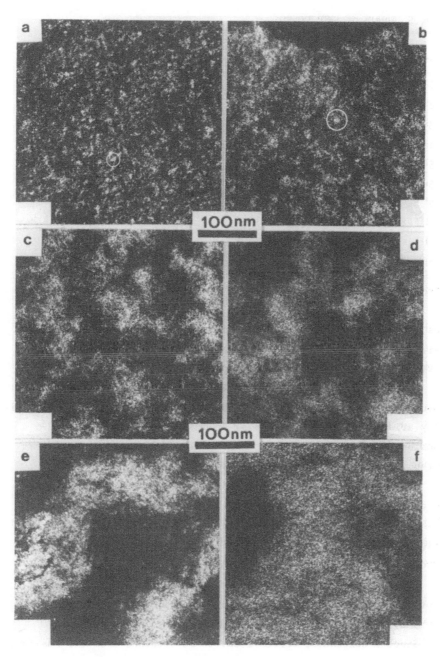

FIG. 61 Various extents of LMO (002 DF): (a) class 2, (b) class 3,
(c) class 5, (d) class 6, (e) class 7, and (f) class 8.

TABLE 4 Chart of LMO Extent

Class	Range of LMO extent (nm)	
1	< 5	
2	5-10	
3	10-15	
4	15-25	
5	25-35	
6	35-50	
7	50-100	
8	> 200	
9	Crumpled lamellae	lamellar textures
10	Flat lamellae	

Source: From Ref. 87.

ones. Figure 62 represents two lamellae seen in BF associated to their schematic representation. Such particles cannot be easily distinguished from flat lamellae when not seen edge on.

The irregular shape of LMO imaged in DF makes it difficult to measure its diameter, i.e., to choose the right class. Another difficulty arises. Most of the samples studied are heterogeneous. To describe LMO accurately, it is thus necessary to obtain a histogram of the frequency of the various classes.

A series of BF and 002 DF taken at a constant low magnification are collected. The displacement of the aperture relative to the 002 ring is chosen so as to explore a whole half circle (Table 2). The surface of the particle projection in the image is evaluated. Then histograms are established from about 100 measures [17,126].

If the model of crumpled sheets of paper described above is adopted, it is possible to derive easily each class of LMO from the others by only decreasing or increasing the crumpling (Fig. 63).

Since a continuous series seems to be obtained from the smallest pores to flat lamellae, a single mechanism has to be assumed [3,14-21, 24,27-30,90,91,116-121]. At first the antagonistic role of oxygen

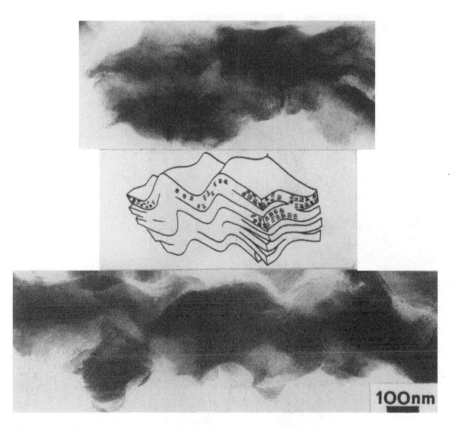

FIG. 62   Sketch of a crumpled lamella and lamellae in BF. (From Refs. 17 and 21.)

and hydrogen was recognized in kerogens, coals, and in oxidized products: the LMO extent was at a minimum for an oxygen-rich mate-rial and at a maximum for a hydrogen-rich one. As a comparison, saccharose-based cokes (SC) belong to class 1 whereas anthracene-based ones (AC) are made of lamellae [16,23]. All intermediates were found in the above-mentioned works, showing an increase in the average LMO extent as $(O/H)_{at}$ decreases. Progressively, quantitative data began to be obtained, which forced authors to be precise about which HTT the elemental analysis was performed. It became apparent that $(O/H)_{at}$ had to be measured as close as possible to the solidification point. This suggests that oxygen atoms still present at the solidifi-

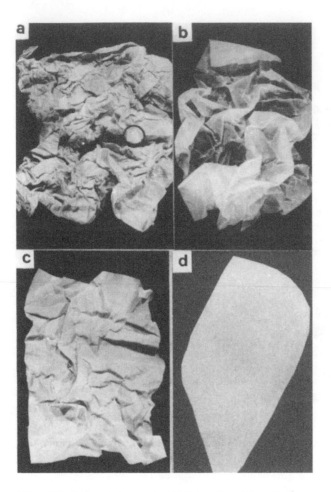

FIG. 63 Decreasing crumpling of a sheet of paper (LMO is circled):
(a and b) pores of increasing radius of curvature, (c) crumpled
lamella, and (d) flat lamella. (From Ref. 121.)

cation point where LMO is definitively stabilized insure a cross-
linking. The authors accepted the earlier explanation proposed by
van Krevelen for coals [127,128], admitted by others [13], and they
generalized it to other carbonaceous materials. Thermal cracking of
carbonaceous matter produces light hydrogenated molecules occurring
simultaneously with the release of tars (aliphatic CH groups). These
molecules act as a suspensive medium for the BSU considered as colloi-
dal micelles. The more numerous the aliphatic CH groups, the larger

the mobility of BSU in the suspensive medium. The more oxygen, the stronger the cross-linking, i.e., the smaller the mobility of BSU. In addition, the more oxygen, the larger the amount of H_2O formed, i.e., the smaller the hydrogen amount remaining. Correspondingly, the minimum viscosity increases and the temperature range of softening narrows (see Fig. 44). The large extent of LMO favored by mobility of BSU becomes small because of cross-linking. This explains well why it is necessary to measure oxygen content at the solidification point. At this point, the hydrogen content corresponds to aromatic CH groups. The role of hydrogen is thus twofold. It is the source of suspensive media for BSU (aliphatic groups), and by saturating (through aromatic groups) the dangling bonds when BSU are free radicals it prevents strong bonding between two adjacent BSU. This dual role favors the formation of large LMO, whereas oxygen limits the LMO extent. This hypothesis well explains why the extremum of aromatic CH (Fig. 43) immediately follows the maximum in the spin concentration measured by RPE (see Section IIIC1).

For samples containing sulfur in addition to oxygen (heavy petroleum products), the interpretation became difficult [19,21,90, 121,129,130]. Cokes obtained from sulfur-rich materials did not obey a simple rule: sometimes sulfur was responsible for a decrease of LMO extent, and sometimes not. The explanation was found by chemical analyses performed on various samples progressively heated from room temperature to 2500°C [130]. Figure 64 shows the behavior of a heavy residue of Boscan heavy oil as compared to light products from light Arabian oil (ALAR) and intermediates such as Athabasca and Safanya. Safanya (triangles) and ALAR (crosses) contain only labile sulfur, which disappears entirely at 1700°C. On the other hand, Boscan (stars) keeps a noticeable amount of sulfur up to 2000°C. For all materials an important release of sulfur occurs between room temperature and 500°C (semicoke or LMO occurrence). Safanya and ALAR cokes have large LMO, whereas Boscan has small ones and Athabasca is intermediate. This demonstrates that only the stable sulfur is responsible for the decrease of LMO extent.

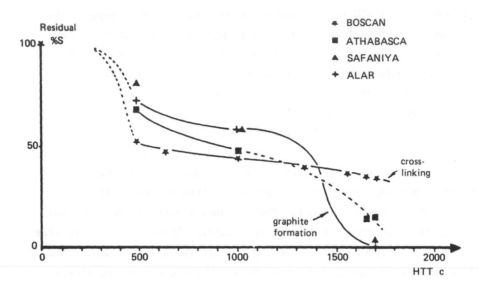

FIG. 64 Sulfur retention in petroleum products (labile and cross-
linking sulfur). ALAR and Safanya contain only labile sulfur; cross-
linking increases from Athabasca to Boscan. (From Refs. 3 and 184.)

Finally, stable sulfur was considered as a cross-linking agent
in a similar manner to oxygen. It was thus necessary to introduce
its content in the $(O/H)_{at}$ ratio measured at the solidification point.
Sulfur cannot be measured at this final stage of LMO: sulfur content
has to be separated into the real cross-linking part and the part
labile before 1700°C, which plays another role. A very simple arithmetic calculation easily converts the amount of S_r (cross-linking sulfur) measured at 1700°C to the temperature of the LMO stage, knowing
the elemental analysis of the precursors, that of the semicoke, and
the weight losses both at the LMO stage and at 1700°C. An important
parameter is thus determined: $F_{LMO} = [(O + S_R)/H]_{at}$, which entirely
governs the LMO extent [90,130]. Figure 65 shows the relation
between F_{LMO} and the extent of LMO evaluated by TEM for classes 1-8
or OM for the medium and coarse mosaics (classes 9 and 10).

   iii. From nucleation to graphitization. Now that the various
extents of LMO marking the end of primary carbonization have been

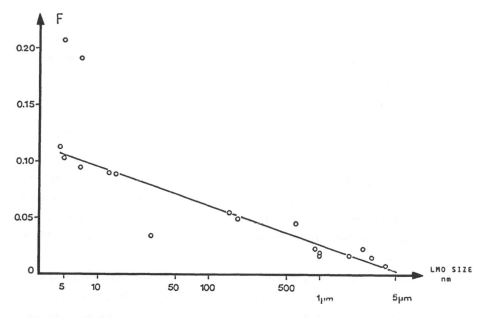

FIG. 65   LMO size versus F_{LMO}.   (From Ref. 90.)

recognized and explained depending on the elemental composition of
the material, it is necessary to establish the affiliation between
the extrema.   It will be shown that a unique mechanism governs the
behavior of any carbonaceous material from nucleation to graphitization.   Then it will be demonstrated that between graphitizing and
nongraphitizing carbons no gap is found but rather an abundant series
of intermediates.

Bonnamy followed not only the carbonization of light petroleum
products but also that of heavier products so as to obtain data for
a continuous and large range of LMO [90,91].   The sample final extents
of LMO were coarse mosaics (class 9) and large LMO (class 8 and 7)
optically isotropic.

When samples were classified by decreasing size of their final
LMO, it was shown first that the same successive stages occur during
carbonization, i.e., nucleation, growth, formation of LMO, then
stabilization of the LMO at the solidification point (end of primary
carbonization).   Then it was shown that:

1. *Coarse mosaics* derive from ellipsoidal nuclei growing into
    Brooks and Taylor single spheres, themselves freely growing
    to a large diameter. These spheres then coalesce forming
    mosaics. The mosaics continue to grow until the solidifica-
    tion point. Nucleation occurs at a relatively high tempera-
    ture ($465°C$ at $4°C$ min^{-1}, $435°C$ at $0.5°C$ min^{-1}) and so does
    solidification (well above $485°C$ and well above $450°C$).

2. *Medium mosaics* derive from the same nuclei but they are more
    dense [compare Fig. 48(b) with (c)]. Therefore single spheres
    cannot grow as much as the preceding ones since they are from
    the beginning very close to each other. They coalesce at a
    lower temperature. Nucleation temperature decreases to $425°C$
    at $0.5°C$ min^{-1}.

3. *Fine mosaics* derive from the same nuclei but no single sphere
    can be formed because the nuclei coalesce before growing into
    Brooks and Taylor spheres. In a thin section the density of
    nuclei is at a maximum as soon as nuclei are visible and they
    touch each other so that they promptly coalesce.

4. *Large LMO* do not derive from ellipsoidal nuclei but from
    nuclei having irregular shapes due to an early coalescence.
    They grow a little and stop. The nuclei when they become
    detectable are not yet solidified, since they are destroyed
    if the material is ground instead of being thin sectioned.
    Figure 66 is an example of such nuclei as seen in 002 DF.
    Their irregular shape is attributed to the fact that they
    coalesce as soon as they are formed. The temperature of
    nucleation is between 420 and $425°C$ at $0.5°C$ min^{-1}.

In summary, the temperature where nuclei occur increases, the
solidification temperature increases also, and the temperature range
of softening enlarges as LMO extent increases. Faster heating rate
[14,29,90,91,137] and higher pressure [14,29] cause the same phenomena (see Fig. 43).

Just before secondary carbonization, in all materials pores
develop by the mechanism described in Section IIIC1. They increase

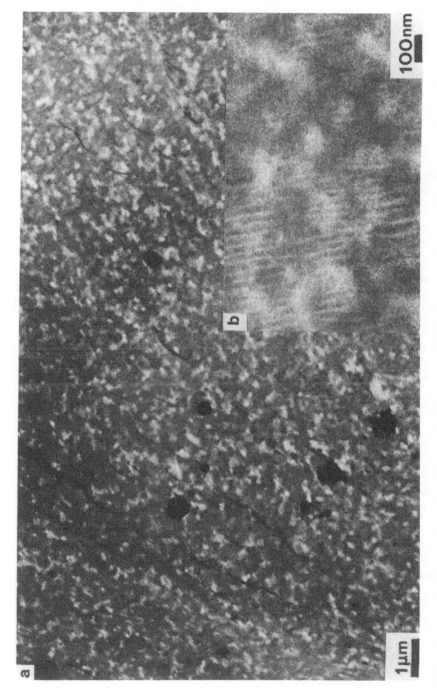

FIG. 66 (a) Nuclei coalesced before solidification (LMO of class 8), 002 DF. (b) Inset: enlargement.

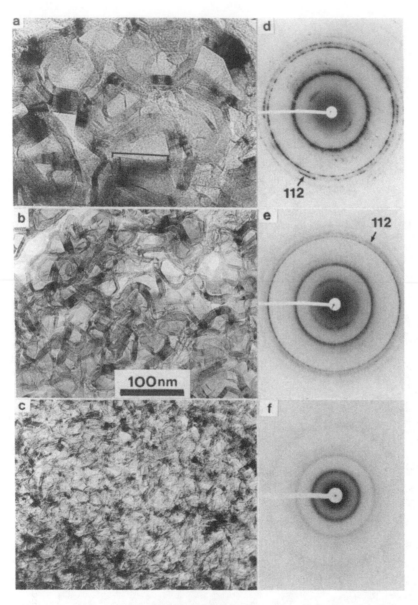

FIG. 67 Cokes heat-treated at 2800°C. Initial LMO size decreases
from the top to the bottom. Formation of polyhedral pores. The
length of their walls corresponds to LMO size. BF (left) and SAD
patterns (right). (a) and (d) Class 7, (b) and (e) class 5, (c) and
(f) class 3. (From Ref. 3.)

in size, i.e., their walls increase in diameter as LMO of the semicoke increases (see Fig. 67).

The above data demonstrate that a continuum exists between the
precursors described and that a unique mechanism governs carboniza-
tion. During primary carbonization nuclei are formed that grow but
may coalesce before reaching the spherical shape. At the end of
primary carbonization a final stage of LMO is reached (solidification,
i.e., semicoke), the extent of which only depends on the atomic ratio
of cross-linking atoms and hydrogen. Coalesced nuclei lead to classes
1-8, early coalesced spheres yield crumpled lamellae (class 9), and
spheres coalesced later give flat lamellae (class 10). There is no
gap between microporous hard carbons and soft carbons but a continuous
series of intermediates whose behavior during cokefaction is entirely
fixed by the size of their "mosaics" formed at the end of primary
carbonization.

The studies carried out upon the graphitization of the various
intermediates described above give an additional argument for the
occurrence of continuous series [3,16-21,30,33,120,121,126,129].
Figure 68 corresponds to 15 carbonaceous materials of different elemental composition heat-treated up to 2900°C. The P and $\overline{d}_{002}$ were
measured by x-ray diffraction. The graphitizability progressively
decreases from Athabasca 2 (sharp line) to R.15.769, whereas the LMO
also progressively decreases from class 10 down to class 2-3 [see
Fig. 67(d,e,f)]. A quasi-linear relationship was found between F_{LMO}
and $\overline{d}_{002}$ [21], emphasized further in unpublished results (Fig. 69).

Finally, it was observed during heat treatment up to 2900°C,
whether partial graphitization occurs or not, that all materials
follow stages similar to these described in Section IIIC1b. BSU
associated with tilt and twist boundaries (left column in Fig. 70)
represent the initial stage (stage 1), then columns and distorted
layers are formed (stages 2 and 3), and partial graphitization takes
place after the stiffening of the layers (stage 4) (right column in
Fig. 70).

Figure 70 should be compared with Fig. 67. Both figures correspond to the same cokes having LMO of decreasing sizes (left column

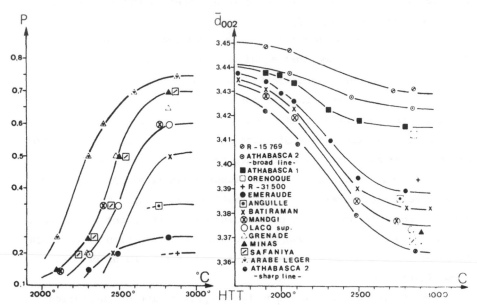

FIG. 68 Decrease of the ability to graphitize (P and $\bar{d}_{002}$) in some petroleum products. (From Ref. 19.)

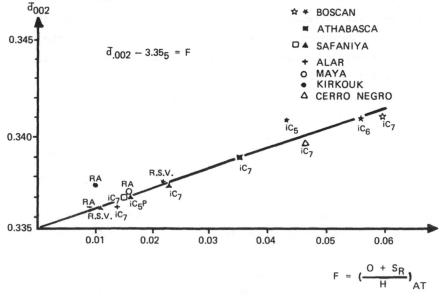

FIG. 69 Variation of $\bar{d}_{002}$ versus $F = [(O + S_R)/H]_{at}$ at the LMO final stage. (From Ref. 21 and S. Bonnamy, unpublished results.)

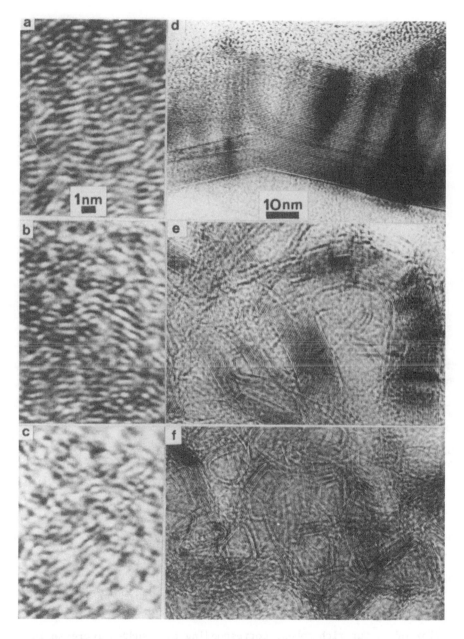

FIG. 70   Same cokes as in Fig. 67.   LMO size decreases from the top
to the bottom.   002 LF.   Left column, cokes; right column, same cokes
heat-treated at 2800°C.

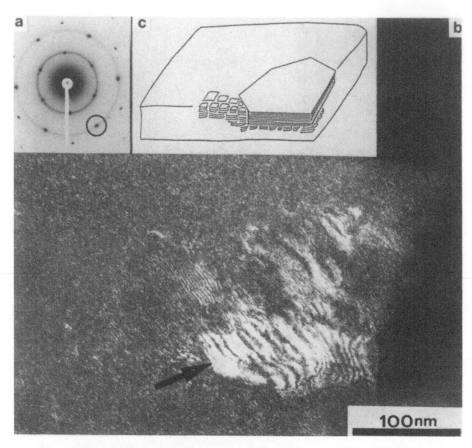

FIG. 71 ALAR coke heat-treated at 1400°C. Labile sulfur as a modifier. (a) SAD pattern (the objective aperture is circled), (b) 11 (0)
DF (the graphite is arrowed), and (c) model.

in Fig. 70) heat-treated at 2800°C (Fig. 67 and right column in Fig. 70).
Whatever the ability to graphitize of the material, the wrinkled layers
of the coke are changed into stiff and perfect layers at 2800°C whereas
graphitizability decreases from Fig. 67(d) to (f).

   iv. Peculiar effects of labile sulfur (puffing). An important
textural change occurring during heat treatment is known as the puffing
behavior of sulfur-rich cokes, corresponding to a sudden increase in
the volume of the coke. This increase takes place after the release
of the labile sulfur. Among the abundant literature devoted to that

phenomenon [130-138], some papers recognize the sudden occurrence of a new phase supposed to be highly graphitized: $\bar{d}_{002}$ is 0.336 nm and a 112 modulation of the 11 band is observed. Both x-rays and TEM were applied by Bourrat et al. [130,137]. X-Ray diffraction using photographic recording (Guinier de Wolff monochromatized chamber) is a valuable tool to evidence the sudden occurrence of a very small amount of crystalline phase among a turbostratic matrix. A sharp reflection at 0.336 nm appears in a sample of Arabian light atmospheric residue (ALAR) heat-treated at 1400°C corresponding to the fast release of sulfur (Fig. 64). At such a temperature the bulk is still turbostratic. Usually, when a specimen of not yet graphitized carbon is dusted onto the supporting film, the lamellar fragments provide SAD patterns only containing hk(0) rings. In some areas hexagonal single crystal patterns superimpose on the 10(0) and 11(0) rings [Fig. 71(a)]. The two components can be imaged simultaneously in DF by setting the aperture on a portion of 11(0) including one of the single-crystal reflections. In the corresponding DF image the turbostratic matrix would be expected to be similar to Fig. 14(a), i.e., weakly illuminated and containing turbostratic moiré fringes (see Section IIB5b). In Fig. 71(b), an additional very bright image full of Bragg fringes is seen associated with the expected one. This image disappears if the aperture is rotated relative to 11(0) so as to avoid the sharp reflection. The very bright area is thus probably responsible for the 0.336-nm spacing detected by x-ray diffraction. Figure 71(c) represents a model of the observed fragment. The assumption that the brilliant image is that of a highly graphitized area is confirmed by ODP of 002 LF images. The 002 interfringe spacing of the matrix (distorted fringes) is larger than the spacing of the single crystal (straight and perfect fringes). Their ratio verifies the numerical value of 0.344/0.335. A rapid exploration in 11 DF of TEM grids containing fragments numerous enough to evaluate the relative proportions of the two phases fits well with the amount of graphitized phase deduced from x-ray patterns. However, the amount of the new phase is proportional not to the initial sulfur content

[139,140] but roughly to the amount of sulfur released between 1300
and 1700°C, that is, labile sulfur (see Fig. 64). This process has
been assumed to be catalytic. In fact, *sulfur is a modifier* since
the process is not at all repetitive. Clearly, sulfur acts only
once when it leaves the carbon. The sudden occurrence of graphite
is not inhibited in the partially graphitizing or even nongraphitizing carbons as Ref. 135 claims. There is only an antagonistic effect
between stable cross-linking sulfur, which decreases the graphitizability, and labile modifier sulfur, creating graphitized areas in
the sample.

2. *Pyrocarbons*

Low- or high-temperature pyrocarbons (PC) illustrate both the tendency
of carbon species to deposit flat on any substrate and the various
statistically spherical microtextures with widespread radii of curvature. A series of reviews has been given [141-145] to which the reader
may refer. This section will be devoted to giving examples of microtexture related to structural and optical properties as compared to
the other kinds of carbons described in Section IIIC1.

All pyrocarbons are prepared by pyrolysis of a hydrocarbon gas
(usually methane) deposited upon various substrates in the temperature
range < 1600°C for low temperature pyrocarbons and 1800-2300°C for
high temperature ones.

a. *Low-Temperature Pyrocarbons*. Their nomenclature was defined from
OM data collected on PC deposited onto fibers. Due to the rotational
symmetry of the fibers, the transverse polished sections of the deposit
also show a rotational symmetry, that is, black crosses similar to
those of Fig. 47(a). The rough laminar texture presents a very strong
optical anisotropy joined to an irregular black cross. The smooth
laminar has a lesser anisotropy but a perfectly smooth cross. The
isotropic is optically isotropic.

These materials [146] are graphitizing carbons for the rough
laminar, and nongraphitizing for the isotropic materials. In fact,
studies of thermal evolution of low-temperature PC followed by x-ray
diffraction and TEM [147,148] show an extreme heterogeneity of indus-

trial samples containing a major phase and the two other species as
minor components. The rough laminar PC are made of lamellae curved
so as to reproduce the so-called "cauliflower" texture of high-
temperature PC [Fig. 72(a)]. Owing to their lamellar texture, they
are characterized by a very high local anisotropy. However, the
radius of curvature of the "cauliflowers" being a few micrometers,
it produces a strong long-scale misorientation responsible for the
rough black cross. Conversely, the smooth laminar PC are formed of
irregular nanometric flattened pores ($\sim$ 100 nm). Their flattening
plane is parallel to the substrate fiber [Fig. 72(b) and (c)]. This
is a much lower local anisotropy than that of rough laminar PC because
the curvature of the layers in smooth laminar PC is more than 10 times
smaller than in rough laminar PC [compare Fig. 72(a) and (b,c)].
Nevertheless, it provides a homogeneous statistical long-range orientation. The black cross becomes smooth only because the radius of
curvature of the pores is beyond the OM resolution. The isotropic
PC are microporous and absolutely similar to Figs. 67(c) and (f) and
70(c) and (f). Their radius of curvature is also 10 times smaller
than that of the smooth laminar PC.

Following the same rule as any other carbons, the rough laminar
PC, having large lamellar LMO, are thus graphitizing. Smooth laminar
PC are only partially graphitizing because the pore walls correspond
to medium-size LMO. Isotropic carbons are nongraphitizing.

b. High-Temperature PC. It is well known from literature that high-
temperature PC have a statistical lamellar texture with growth cone
features observed on a large scale by OM. The cone axes are parallel
to the growth direction of the deposit (cauliflower-like texture).
They are graphitizing carbons but follow very different kinetics compared to other carbons [149-151]. They remain turbostratic up to
2600°C, then graphitize rapidly. Goma et al. [35,148], by comparing
TEM and x-ray data have shown very peculiar features in the microtexture of 2100°C as-deposited PC. The 002 LF show, that the layers
of the deposit are straight and perfect as in all other carbons heat-
treated at this temperature [compare Fig. 73(a) to Fig. 57(b) and

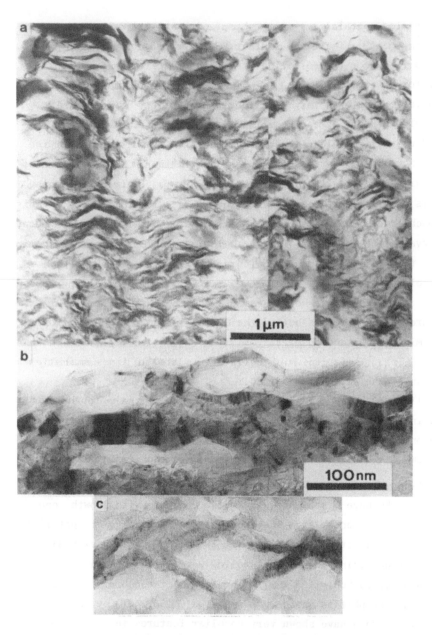

FIG. 72 Transverse thin sections of low-temperature PC deposited upon a carbon fiber (the fiber substrate is far below the bottom of the figure). (a) Rough laminar PC. (b and c) Smooth laminar PC [isotropic PC is similar to Fig. 67(c,f)]. (From Ref. 147.)

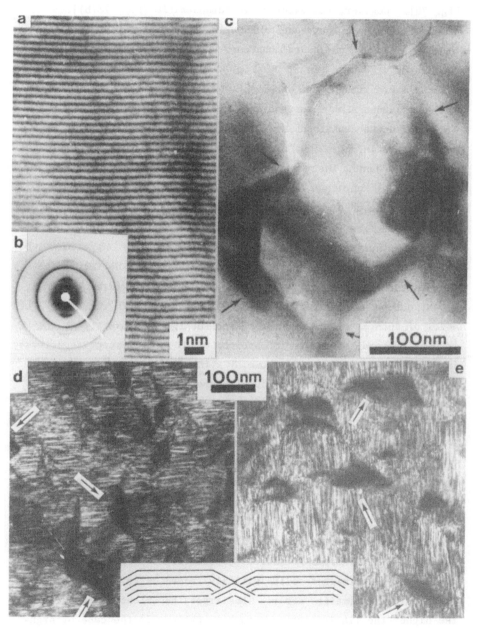

FIG. 73 2100°C As-deposited PC: (a) 002 LF, (b) SAD pattern, (c) BF,
and (d and e) orthogonal 002 DF. Inset, model. (From Ref. 35.)

Fig. 70]. However, the material is turbostratic [Fig. 73(b)].
Samples containing the plane of deposition were prepared by ion
thinning. In BF, polyhedral contours due to diffraction contrast
appear [Fig. 73(c)]. In 11 DF [Fig. 73(d) and (e)] rotational turbostratic moiré fringes are observed [see Fig. 14(a) and Section IIB5b]
with dark areas always perpendicular to the moiré fringes. By tilting
the specimen as explained in Section IIB5b and by applying the same
reasoning as in Fig. 27(b) and Section IIIA1b, the thinned sample was
shown to be bent. Therefore, the extinction contours appearing dark
in Fig. 73(b) represent the oblique areas of the specimen. Their
angle with the observation plane can be deduced from Table 2, Fig. 19
and aperture size (± 27° in the case of Ref. 35). The radius of curvature can thus be calculated to be about 50 nm. The shape cannot be
smooth but polyhedral since the layers are stiff and perfect. The
microtexture is therefore cauliflower, as was the texture observed
by OM, but on a much smaller scale ("microcauliflower" texture). The
model of such a texture is represented in Fig. 73 (inset). As the PC
is heat-treated, no change is observed up to 2600°C except a decrease
in $\overline{d}_{002}$ down to 0.338 nm as shown by x-ray diffraction and TEM.

Above 2600°C, TEM shows the disappearance of the "microcauliflower" texture. It is replaced by planar domains separated by
classical grain boundaries. The transformation occurs at random in
the specimen so that the material is heterogeneous. Moreover, if the
cauliflower-like areas are always turbostratic whatever the temperature, planar ones are graphitized with various degrees. These planar
areas appear above 2600°C and increase in number to reach almost 100%
at 2800°C HTT. TEM data may explain the mechanism of this sudden
graphitization. Let us consider first the thermal progressive graphitization. It is normally ensured (Section IIIA1b) by successive elim-
ination of defects in a solid initially formed of very small BSU
arranged with "nematic" order. The larger the oriented zones (LMO),
the higher the final degree of graphitization. In the case of high-
temperature PC, on the contrary, such a progressive improvement is
not possible. This is due first to the layers being already perfect,
and second to the large-scale distortions being highly energetic

because of the polyhedral shape of "microcauliflowers." However, shear stresses develop in such a product, increasing with HTT. Above a defined temperature the development of these stresses causes the breakage of the boundaries, i.e., partially graphitized areas are suddenly produced in increasing number. Nevertheless, a lower final value of P can be expected and is observed: P_{max} = 0.7 [35]. The same mechanism of sudden changes will be described in the next section for nongraphitizing carbons heat-treated under pressure. It was already cited in literature that incomplete graphitization of PC could be achieved above 3200°C [150]. However, at that temperature the vapor pressure of carbon is high enough to give spurious effects due to vapor growth or vapor etching. The sublimation studies give a vapor pressure of 1 bar at about 3600°C [152,153].

As usual, properties other than structural ones depend on microtexture changes. Magnetic susceptibility, for instance, shows a very strong change of slope versus HTT above 2500°C (Fig. 74) before reaching the value of graphite above 3200°C. It is clear that the minimum observed is due to the sudden flattening of a maximum of "microcauliflowers" themselves. Resistivity in the direction of the layers shows a sudden decrease at 2600°C, which is coherent with the flattening of

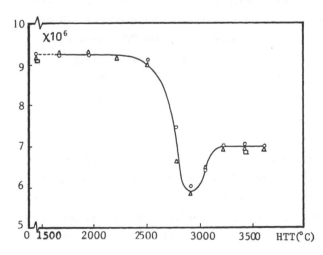

FIG. 74 Magnetic susceptibility of high-temperature pyrocarbons of various origins. (From Ref. 149.)

the "microcauliflowers." Magnetoresistance also shows a disconti-
nuity at 2600°C.

3. *Graphitization under Pressure*

The previous data (Sections IIIA2c and IIIC2b) show a strong tendency
of carbons having gained stiff and perfect layers too early, or in
an "abnormal" way, to be thermally stable as far as their ability to
graphitize is considered. Such a feature is also observed in hard
carbons heat-treated under pressure. In the same way as high-
temperature PC they graphitize by sudden changes applied in areas
whose number increases. These areas are distributed at random within
the specimen. Various examples were studied by x-rays and TEM [154-
160] among industrial nongraphitizing carbons and natural samples
(anthracites, graphitoids, semigraphites). These studies followed
an important series of experiments carried out under pressure by
Inagaki et al. [161]. These authors showed nongraphitizing carbons
to be suddenly transformed into graphite under a pressure of 5 kbar,
above 1600°C. The example given here is that of a glassy carbon.
X-Ray diffraction profiles of 002 show a first sudden change at
1300°C corresponding to the development of a peak at 0.339 nm and
a second one above 1600°C for which $\overline{d}_{002}$ is 0.336 nm (Fig. 75). At
that stage 10 and 11 split into sharp hkl reflections. The initial
samples are microporous, i.e., similar to the bottom of Fig. 70 or
even with smaller or much smaller pores. From 1100°C hollow entangled shells appear (Fig. 76). Their shape is polyhedral [Fig. 76(a)].
Their walls contain perfectly straight fringes [Fig. 76(b)]. They
appear at random in the specimen and their number increases up to
1600°C, where they make up almost the whole sample. Their microtexture was precisely determined by applying to hk DF images the
same reasoning as in Section IIIA1b. Data show microtexture of the
hollow shells to be that of crumpled sheets of paper (see model in
Fig. 63). It is also that of the glassy carbon, although much more
crumpled. Obviously, the hollow shells derive from the micropores
by a sudden and large increase in the radius of crumpling of the
sheets. Between 1600 and 1700°C all the shells break into highly

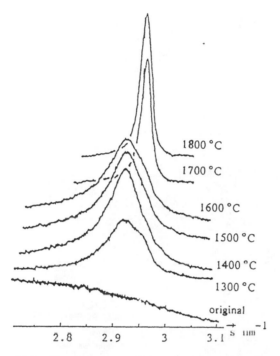

FIG. 75 X-Ray profile of 002 reflections. Glassy carbon heat-
treated under 5 kbar. (From Ref. 156.)

graphitized lamellae. The same stage of transformation were observed
in anthracites [154] and in many other natural carbons [159,160].

If we consider the regions where the crumpled sheets of glassy
carbon layers touch each other and if grain boundaries happen to
coincide at one point, bonding occurs. Indeed, such boundaries contain many active defects such as dangling bonds, heteroatoms, sp^3
bonds, etc. [152,153]. In such a complex texture, shear stresses
develop under pressure. Even when submitted to isostatic pressure,
the resultant of the stresses is not zero when the material is
porous. They concentrate at the boundaries. Thus they cause the
breakage of intersheet bondings and a sudden relaxation of the
corresponding crumpled sheets. The radius of curvature increases
and locally the material could be described by the model of Fig. 63
where the model of Fig. 63(a) would suddenly explode into that of

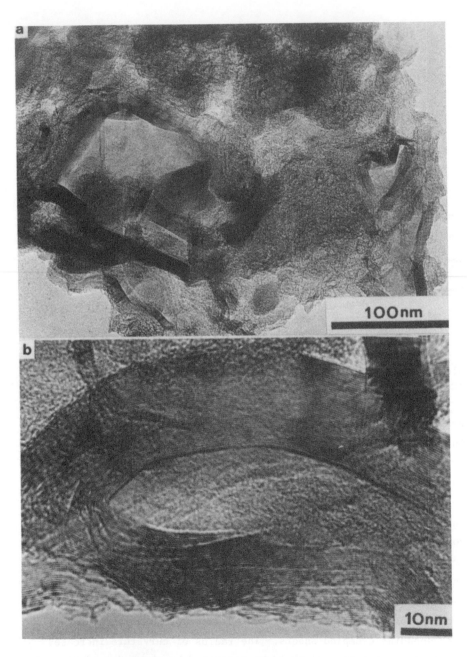

FIG. 76 Glassy carbon heat-treated under 5 kbar (1300°C): (a) BF
and (b) 002 LF. (From Ref. 156.)

Fig. 63(b) in the area where stresses are high enough. It is exactly the mechanism of relaxation of the distortions in PC already described. This mechanism is also followed in natural graphitization, which always occurs under stresses [154,158-160].

IV. CYLINDRICAL CONFIGURATIONS

As in Section III (spherical symmetry), two kinds of materials can be distinguished: the first ones belong to the true cylindrical symmetry, among which are the filamentous carbons, and the second ones belonging to statistical symmetry (PAN-based or pitch-based fibers).

Except that a driving force directed along an axis exerts its influence during carbonization and/or carbon growth, the behavior of these carbons follows the same rules as that listed previously. Their properties (particularly mechanical and electrical properties) also depend on their microtexture.

A. Filamentous Carbons

Many kinds of chemical reactions producing carbon are known, initiated by a catalyst. The Boudouard equilibrium (disproportionation) $2CO \rightarrow CO_2 + C$ is catalyzed by Fe, Ni, Co, and their alloys. Other gases or vapors (methane, benzene) are also able to decompose at the contact of a catalyst at various temperatures. The literature is so abundant on that subject that all references cannot be quoted here. They will be found in reviews or theses [61,67,162-163]. The examples given here will be restricted to hollow fibers [Fig. 77(a)] grown in vapor phase from benzene [25,26,163] and to carbon formed from CO disproportionation [67,68].

1. Hollow Fibers

Such fibers were prepared by Koyama et al. [164] by pyrolyzing benzene diluted in hydrogen at about 1100°C. The fibers grow in two steps. First, a very thin hollow catalytic filament is formed from microcrystals of iron oxide reduced into iron by hydrogen. Then an increasingly thick pyrocarbon deposit is formed on that substrate. The microtexture can be determined only by DF and 002 LF on trans-

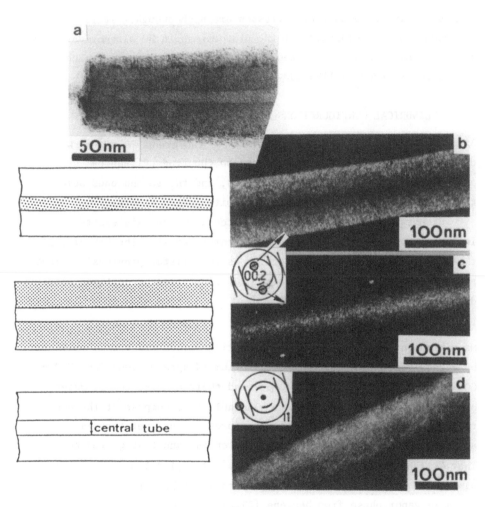

FIG. 77 True cylindrical texture with a hollow core, sketches and
images: (a) BF, (b) 002 DF, (c) 10 DF, and (d) 11 DF (compare with
Fig. 21). (From Ref. 25.)

verse thin sections, and by longitudinal observations. The cylindrical texture [Fig. 1(f)] was proved [25] by combining 002 DF [Fig.
77(b)], 10 DF [Fig. 77(c)], and 11 DF [Fig. 77(d) in the same manner
as in Section IIIA1b, whereas BF [Fig. 77(a)] demonstrates the occurrence of a central hollow channel. By examining 002 LF micrographs,
two kinds of carbon layers were found (Fig. 78). Near the core are

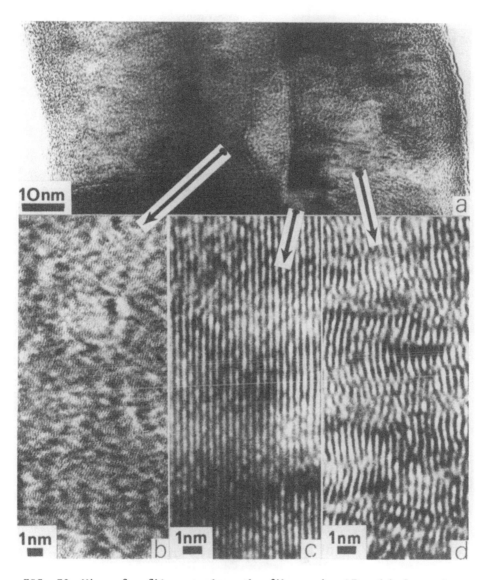

FIG. 78 View of a filament along the fiber axis, LF: (a) General
view, (b) 10 LF, (c) 002 LF near the core (catalytic carbon), and
(d) 002 LF of the pyrocarbon deposit. (From Ref. 25.)

perfect, long and straight fringes in stacks of 10 to 30. Zigzag distorted fringes fill the remaining part of the fiber wall (20-50 nm).
Often BF images show broken fibers with a thin hollow core emerging

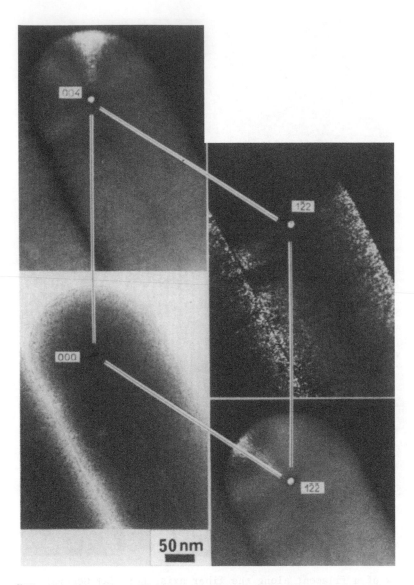

FIG. 79   BF and DF of the tip of a carbon filament including a
crystal of cementite, arranged in the crystallographic positions
given by the illumination of the crystal and the aperture center
position in SAD mode.   (From Ref. 25.)

from the fracture. Often also very thin filaments devoid of pyrocarbon deposit can be found. The catalytic nature of the primary
filament growth was also demonstrated by TEM. In BF images of the
tip of a fiber an unknown dark particle, less than 10 nm in size,
was found encapsulated in carbon (Fig. 79). A complete exploration
of the SAD pattern (see Section IIB4) with a very small aperture,
including numerous overlappings of the aperture opening, allowed one
to determine a planar unit cell deduced from DF images (Fig. 79).
Its angles and the side lengths were measured by considering the
aperture center positions in SAD mode when the particle is illuminated (see Section IIB4). Repeating the same experiments on a series
of similar particles permitted us to identify iron carbide (cementite).

Because the pyrolytic deposit is thick relative to the catalytic
carbon, fibers as a whole are graphitizing carbon [26]. As a result,
mechanical and electrical properties approach those of graphite
whiskers [163].

2. Filamentous Carbons [67,68,165-168]

Disproportionation of CO yields at a low temperature (450-650°C)
tubular filaments or shells of carbon containing a metal microcrystal
less than 100 nm in size [162,169-173]. The work of Audier et al.
[67,68,165-167] emphasizes the role of the crystallographic orientation of the metal microcrystal upon both growth and graphitization
degree of the carbon formed. By combining BF, SAD patterns, DF, and
002 LF using a goniometer stage, it was shown that body-centered cubic
metal catalysts such as FeCo alloys [Fig. 80(a)] favored carbon growth
along their |100| axis [Fig. 80(b)] whereas face-centered cubic cata-
lysts (Fe Ni alloys) were operative along |110|. Correspondingly, in
conical metal crystals the faces free of carbon are (100) for bcc
structures and (111) for fcc structures. In addition, the catalyst
was showed to be often twinned and the twin law determined [166,167].
In the same manner, biconic metal crystals are present in large proportions [167]. As an example, fcc FeNi alloys will be presented.

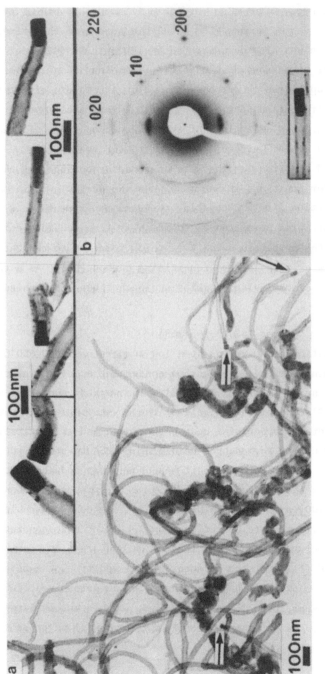

FIG. 80   (a) BF images (and in insets) of bcc Fe-Co alloy crystals and filamentous carbon.   (b) SAD pattern.   (From Ref. 165.)

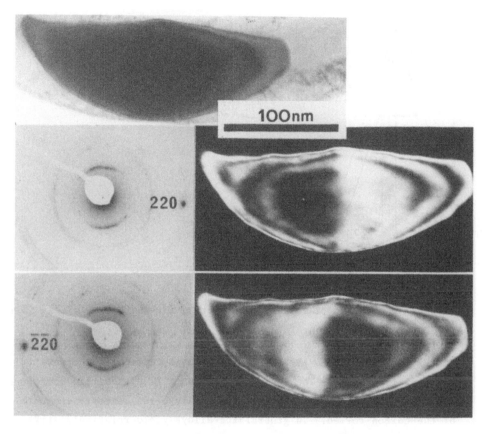

FIG. 81 BF and DF images corresponding to 220 and $\bar{2}\bar{2}0$ associated with the SAD patterns. (From Ref. 167.)

They are formed by the association of two single cones bounded by their (110) face (Figs. 81 and 82). The complete demonstration was given by considering DF images and SAD patterns of a bicone free of twins, oriented with its axis perpendicular to the tilt axis of the goniometer stage. DF images were then taken with 220 and $\bar{2}\bar{2}0$ diffraction spots (see Section IIB4b). Because two opposite areas of the bicone appear successively bright with these two opposite reflections, there are two combined single crystals. Their angle is related to the difference of tilt angle of the goniometer and to the 220 Bragg angle, that is, 3.66°. The only crystallographic possibility to obtain 220

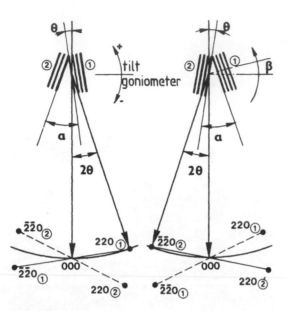

FIG. 82   Sketches of the Ewald construction for the two DF of Fig. 81.
(From Ref. 167.)

and $\overline{2}\overline{2}0$ as indicated by DF and SAD is to associate the two crystals
by their (110) face.   The interface is thus a (110) subgrain boundary.
The crystallographic structure of the catalyst influences the nature
of the carbon formed.   In general fcc alloys yield relatively stiff
and perfect aromatic layers, whereas FeCo gives less perfect layers.
In addition the low temperature carbons produced [68] are often
turbostratic or only partially graphitized in spite of the perfection
of their layers.   All of them are thermally stable, whatever the HTT.
The thickest filaments and the shells (HTT > 500°C) are partially
graphitized, the others (HTT < 500°C) are turbostratic.   Another
demonstration is therefore given that progressive graphitization
cannot occur without initial small defective units oriented in
parallel in large areas.

B.   Carbon Fibers

Production of PAN-based carbon fibers or pitch-based ones is very important for the composites industry.   The understanding of the origin

of their peculiar mechanical and electrical properties is highly
important. A very recent review was given in this series [174].
Another review is mainly devoted to the intrinsic properties and
numerical relations between microtexture and mechanical and electrical
properties [34,175]. These papers discuss also the various models
given for PAN-based carbon fibers. Other papers deal with PAN-based
fibers [11,176,177], pitch-based fibers [178], and heat treatment of
PAN-based high-tensile-strength fibers [179]. Other anterior reviews
[180,181] also have to be cited. It would be too long to account for
the tremendous amount of data gathered in these reviews or papers.
Only a brief summary of the microtexture of high-tensile-strength and
high-modulus PAN-based fibers will be recalled here under the form of
photomontage (see Figs. 83 and 85).

1. High-Tensile-Strength Fibers

These are obviously deriving from carbons having statistical spherical
symmetry by introducing in the crumpled-sheet of paper models (see
Fig. 63 and Section IIIC) an elongation parallel to an axis (fiber
axis). They are also comparable to low-temperature carbons (cokes
obtained during secondary carbonization) as far as their microtexture
is concerned (see Section IIIC and Fig. 63).

Figure 83 shows the model obtained from TEM studies, including
real 002 LF micrographs. The BSU are here similar to those of other
carbons and their tilt and twist boundaries are also comparable to
what is observed during secondary carbonization, i.e., stage 1 of
Fig. 55 (see Section IIIC1b). The important result obtained is the
possibility of classifying commercial samples by increasing close
packing of the crumpled layers, i.e., by decreasing radius of curvature of the sheets either transverse (r_t) or longitudinal (r_ℓ) (see
Fig. 63). Such a classification is the basis of the numerical relationship between microtexture and mechanical properties, e.g., tensile
strength σ_c [Fig. 84(a)]. The physical meaning of the increase of σ_c
with Δ_t is as follows. The distorted sheets of BSU associated with
tilt and twist boundaries are bounded to each other each time two
boundaries of adjacent sheets can get into contact, owing to the many

FIG. 83 Model of high-tensile-strength PAN-based fiber associated
with 002 LF. (From Ref. 176.)

dangling bonds or heteroatoms gathered there. The more compact the
fiber, the higher the chance for adjacent sheets to bind. The lateral
cohesion is thus the cause of the high strength. More precise data
are obtained from high-modulus fibers.

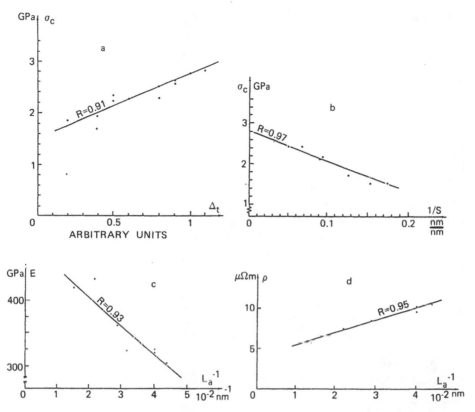

FIG. 84 Numerical relations between some properties and the micro-texture. (a) High-tensile-strength fiber σ_c. (b,c,d) High-modulus fibers, (b) tensile strength σ_c, (c) Young's modulus E_m, and (d) electrical resistivity ρ. (From Ref. 11.)

2. *High-Modulus Fibers*

High-modulus fibers correspond to a heat treatment of high-tensile-strength fibers [179] and are comparable to high-temperature carbons (see Figs. 16, 17, 19, 56, and 57; see also Sections IIB5 and IIIC). Their scattering domains are large and well defined so that $L_{a \text{ll Cr}}$ can be measured easily (see Table 3) from the observed moiré fringes. Correspondingly r_t and r_ℓ are also measurable from 002 LF. Figure 85 corresponds to the same type of photomontage as in Fig. 83. The lateral cohesion of the fiber is also ensured by bonding between

FIG. 85 Model of high-modulus PAN-based fiber associated with 002 LF.
(From Ref. 177.)

adjacent distorted sheets of carbon (see Fig. 19) at the condition
where two grain boundaries collapse. The chances of this to occur
increase as r_t and r_ℓ decrease whereas L_a increases. Hence the
lateral cohesion will be represented by the variable $S = \overline{L}_a[(1/\overline{r}_t) +
(1/\overline{r}_\ell)]$. A linear correlation is observed between σ_c and $1/S$ [Fig.
84(b)]. Young's modulus E is correlated to $\overline{L}_a$, the layer diameter,

as seen in Fig. 84(c), whereas strain ε is proportional to $[1/\cos(Z/2)]\ [(1/\overline{r}_t) + (L/\overline{r}_\ell)]$ where Z is the average misorientation of the layers parallel to the fiber axis. Electrical resistivity itself is correlated to $1/\overline{L}_a$ [Fig. 84(d)].

V. CONCLUSION

The first general conclusion that we can draw from the data gathered in this chapter is the absolute necessity of determining the three-dimensional arrangement of aromatic ring structures that constitute the microtexture of carbons, because all their physical properties depend on it. Examples were given showing the high dependence between microtexture and structural, optical, electronic, electrical, and mechanical properties. For obtaining a reliable model of the microtexture, all possible imaging techniques have to be employed at any possible scale. OM studies are thus complementary to TEM ones.

The second conclusion to be drawn concerns the mechanism of growth of carbons. Two very simple rules are obtained from the points discussed. Aromatic units (BSU) associate edge to edge and face to face so as to form a nucleus. This is found to be the case for mesophages of all kinds (pp. 65-106). The final size and shape of the anisotropic domains forming carbons only depend on the antagonistic heteroatoms such as H versus cross-linkers such as O or S. However the influence of the aromaticity of the initial organic matter was masked by the fact that an increase of aromaticity was often accompanied by a huge increase in the cross linking atom content. The example of radial textures (pp. 55-65) and more recent examples of new textures tend to emphasize the utmost importance of this parameter. Another common type of carbon growth is due to the anisotropy of BSU which tend strongly to lie flat on any available substrate, to form parallel deposits of increasing thickness. This is found to be the case of concentric textures (pp. 32-55) and pyrocarbons (pp. 106-112). Whatever the type of growth, it is always governed by "wettability" rules of the aromatic layers relative to the matrix in or upon which carbons grow.

VI. APPENDIX

A. Introduction

The reciprocal nodes of a crystal lattice are elongated in the direction where the crystal is the thinnest. Their resulting shape is
thus entirely dependent on the crystal shape. A thin platelet having
a very large diameter will give spikes, whereas a thick platelet will
give small dots. In the case of a material devoid of stacking order
(two-dimensional crystal), the crystal is reduced to one lattice plane
and the reciprocal nodes are continuous lines.

   The shape of the reciprocal node is given by the Fourier transform of the crystal shape (Fig. 86). In the case of this figure, the
node can be approximated to a spike having a length 1/t if the crystal
thickness is t.

   When a node is elongated enough, it can touch the Ewald sphere
even if the corresponding lattice planes are exactly parallel to the
incident beam (Fig. 87). The hkl and $\overline{hk\overline{l}}$ reflections thus occur
simultaneously. In this case, the node elongation corresponds to a
crystal thickness $t = d/2\theta = d^2/\lambda$ (Fig. 87). As an example, for the

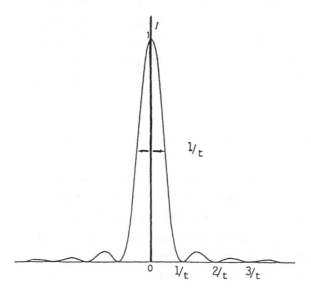

FIG. 86 Shape of a reciprocal node.

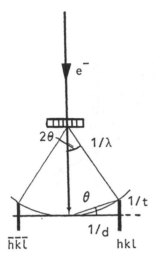

FIG. 87 Ewald construction for a thin crystal.

002 reflection of carbons, t must be less than or equal to 34 nm for 120 kV. Such ranges of thickness are practically the rule in all carbons except when most highly graphitized.

B. Reciprocal Space of Turbostratic Layer Stacks Oriented Parallel

The reciprocal space of a single layer having a diameter L_a is a hexagonal lattice where nodes are replaced by reciprocal infinite hk lines [Fig. 88(a)]. The transverse profile of each line is directly connected to the diameter L_a of the layers. The longitudinal profile follows the scattering atomic factor of carbon. The intensity is thus regularly decreasing along the Z ordinate of the reciprocal space. It is maximum for Z = 0. If two layers are piled up in parallel with a small rotation α, they give a reciprocal space deduced from Fig. 88(a) by a rotation of α [Fig. 88(b)]. The 00ℓ nodes are moreover produced. Their maximum order depends on the perfection of the layer parallelism. Figure 88(b) represents the reciprocal space of a single turbostratic stack. Now, if the material is made of a great number of such stacks in azimuthal disorder, the final reciprocal space is obtained from Fig. 88(b) rotated around the OZ axis. The hk concentric cylinders are obtained in addition to 00ℓ nodes along OZ (Fig. 89). Such a kind

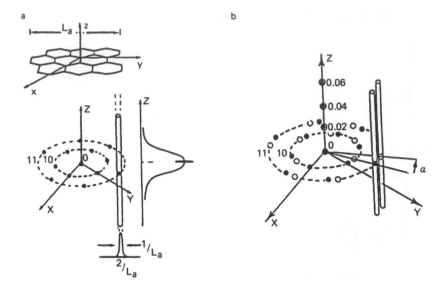

FIG. 88 (a) Reciprocal space of a single aromatic layer.
(b) Reciprocal space of a turbostratic layer stack. (From
Ref. 33.)

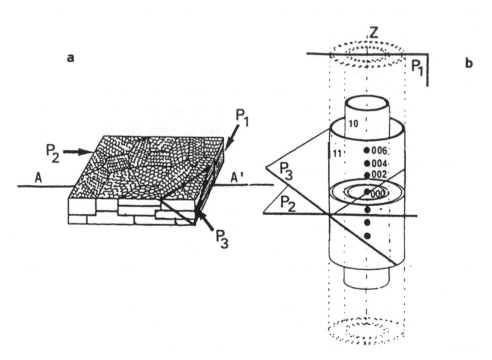

FIG. 89 (a) Turbostratic stacks in azimuthal disorder. (b) Reciprocal
space. (From Ref. 33.)

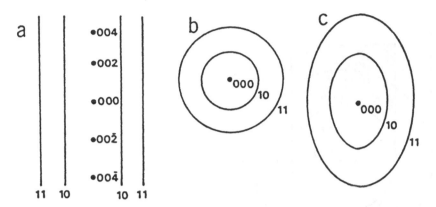

FIG. 90 Sketches of SAD patterns issued from Fig. 89. (From Ref. 33.)

of disorder is very often encountered in soft carbons where the LMO is large.

Considering Fig. 89, the SAD patterns obtained depend on the orientation of the particle relative to the incident beam. We may distinguish three different orientations, along planes P_1, P_2, and P_3. P_1 corresponds to a particle seen edge-on. This reciprocal space section is sketched in Fig. 90(a). P_2 corresponds to a particle lying flat on the supporting film. The reciprocal space is thus a section along P_2 [Fig. 90(b)]. The third oblique possibility is P_3 [Fig. 90(c)]. Figure 90 represents the possible SAD patterns corresponding to sections along P_1, P_2, and P_3.

Some examples are illustrated in Fig. 58 (sections along P_1).

C. Reciprocal Space of a Fiber

A fiber is obtained by rotating Fig. 89(a) around a fiber axis AA'. If we imagine the reciprocal cylinders of Fig. 89(b) rotating continuously around the fiber axis, we obtain concentric 10 and 11 hollow spheres. Their walls are thin along AA' and are asymmetrical along the normal to AA'. In the pattern, hollow circles are obtained with a band profile along the meridian direction. The intensity decreases slowly toward large angles (Fig. 91). The 00ℓ nodes themselves describe circles, which in the pattern give sharp 00ℓ reflections (Fig. 91). In the external part of the pattern the hk lines are

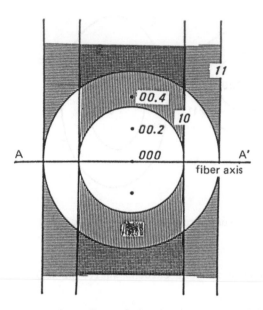

FIG. 91 Sketch of SAD pattern of a fiber.

elongated normal to AA' and show their actual intensity profile. In
fact, the layers are always noticeably misoriented along the fiber
axis AA' so that Fig. 89 should not only be rotated around AA' but
also tilted relative to AA' along the conical axis OZ with a misorientation angle Z. The 00ℓ reflections are thus spread into arcs.
Some kind of limited band effect due to the tilt appears on the hk
line along the equator, whereas the tilt of hk lines becomes more
apparent in the external parts of the pattern.

D. Reciprocal Space of a Complete Random Sample

Now the sketch may be complicated again by the occurrence of random
distribution of the stacks. Figure 89 not only turns around AA' but
also around an axis perpendicular to the plane of Fig. 89. The resulting pattern (Fig. 92) consists of concentric hollow spheres. All
of them except the 00ℓ ones have asymmetrical walls. A band profile
is thus obtained in any direction from the center, except for 00ℓ,
which are symmetrical. The half-width of 00ℓ increases as L_c
decreases. SAD patterns obtained are similar to x-ray diffraction
ones [see Fig. 67(f)].

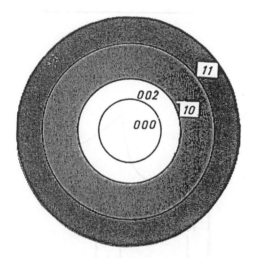

FIG. 92   Sketch of SAD powder pattern.

E.   Graphitization

Graphitization is a statistically homogeneous process due to an
increasing number of pairs of layers acquiring the sequence of
graphite.   This means that they acquire suddenly the 0.335_4-nm interlayer spacing and the AB sequency of graphite.   The average $\overline{d}_{002}$
decreases during the increase of P, the probability of finding
graphite pairs of layers.   Many authors [109,150,182] have shown
that P is not connected to $\overline{d}_{002}$ by a univocal relation and numerous
different relations have been proposed.   It is therefore impossible
to approximate the graphitization degree P by the measurement of
$\overline{d}_{002}$.   Another reason to avoid correlating P with $\overline{d}_{002}$ is the difficulty of taking into account the cases where $\overline{d}_{002}$ is larger than 0.344
nm, the interlayer spacing of a turbostratic carbon.   Moreover, it
is known that many carbons have small $\overline{d}_{002}$ without being either
graphitized or graphitizing carbons (see Sections IIIC2, IIIC3,
and IV).

For measuring reliable graphitization degrees, P has to be
measured.   It is only possible to do this by measuring the intensi-
ties along 11 band in x-ray diffraction patterns [108,111,183].   The
profile of 11 along Z is continuous; it approximates the square of

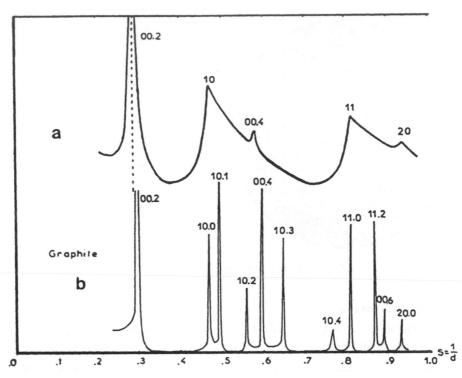

FIG. 93 X-Ray pattern of carbons: (a) turbostratic and (b) graphite.
(From Ref. 33.)

the carbon atomic scattering factor. During the increase of P, the
transition between two-dimensional and three-dimensional cyrstalline
structure causes a progressive modulation of the 11 band into 110 and
112 sharp reflections (Fig. 93). The progressive modulation of hk
bands cannot be measured from SAD patterns, but it can be recognized
and followed (see Fig. 58). In SAD patterns corresponding to P_1
sections, no 00ℓ reflections are present. Hence the advantage of
SAD relative to x-ray diffraction is to avoid any superimposition
between hk lines and 00ℓ reflections whereas 00ℓ is superimposed on
the tails of hk bands in x-ray patterns. The progress of graphitization can also be followed in the oblique SAD patterns along plane P_3
[Fig. 90(c)] but less accurately, because in such oblique sections
each hk (Z) area extends over more than one hk line.

In the case of fiber patterns, graphitization, if any, can be evidenced by considering the external part of the hk cylinder along a meridian direction.

F. Conclusion

In the case of azimuthal misorientation, a series of numerical data can be obtained from SAD patterns. From those corresponding to a section along P_1, a radial photometric recording will yield the width of 00ℓ reflections, that is, $L'_{c\ 00\ell}$ (Table 3). From patterns corresponding to P_2, $L'_{a\ hk(0)}$ will be obtained (Table 3) corresponding to the zero ordinate. From patterns along P_1 and P_3 the graphitization process may be followed (see Fig. 58).

In the case of a fiber, an additional numerical value can be obtained: the opening Z of the 00ℓ arcs, related to the misorientation of the layers relative to the fiber axis.

REFERENCES

1. R. D. Heidenreich, N. M. Hess, and L. L. Ban. *J. Appl. Crystallogr., 1*:1 (1968).

2. Y. Hishiyama, A. Yoshida, and M. Inagaki. *Carbon 20*:79 (1982).

3. A. Oberlin. *Carbon 22*:521 (1984).

4. G. R. Millward and D. A. Jefferson. In *Chemistry and Physics of Carbon*, Vol. 14 (P. L. Walker Jr. and P. Thrower, eds.). Marcel Dekker, New York, 1978, p. 1.

5. X. Bourrat. Contribution à l'étude de la croissance du carbone en phase vapeur. Thèse d'Etat, Université de Pau, France, 7 septembre 1987.

6. A. Oberlin. *Carbon 17*:7 (1979).

7. P. B. Hirsch, A. Howie, R. B. Nicholson, D. W. Pashley, and M. J. Whelan. In *Electron Microscopy of Thin Crystals*. Butterworths, London, 1965, p. 295.

8. C. Fert. In *Traité de microscopie électronique* (C. Magnan, ed.). Hermann, Paris, 1961, p. 367.

9. K. J. Hanszen. In *Advances in Optical and Electron Microscopy* (R. Barer and V. E. Cosslett, eds.). Academic Press, New York, 1971, p. 68.

10. A. Howie, O. L. Krivanek, and M. L. Rudee. *Phil. Mag. 27*:235 (1972).

11. M. Guigon. Relations entre la microtexture et les propriétés mécaniques des fibres de carbone ex-PAN. Thèse d'Etat, Université de Technologie de Compiègne, France, 13 novembre 1985.

12. D. W. Van Krevelen. In *Coal*. Elsevier, Amsterdam, 1961, p. 445.

13. R. Loison, P. Foch, and A. Boyer. In *Le Coke*. Dunod, Paris, 1970, p. 73.

14. A. Oberlin, J. L. Boulmier, and M. Villey. In *Kerogen* (B. Durand, ed.). Technip, Paris, 1980, p. 191.

15. P. Ollivier. Etude de brais de pétrole par RMN et MET. Thèse d'Ingénieur-Docteur, Université d'Orléans, France, 18 janvier 1985.

16. J. N. Rouzaud. Relations entre la microtexture et les propriétés des charbons. Thèse d'Etat, Université d'Orléans, France, 24 avril 1984.

17. F. Bensaid and A. Oberlin. *J. Chem. Phys. 84*:1457 (1987).

18. D. Auguié. Mésophase carbonée. Interaction entre coke de pétrole et brai de houille. Thèse d'Ingénieur-Docteur, Université d'Orléans, France, 10 décembre 1979.

19. M. Monthioux, M. Oberlin, A. Oberlin, and X. Bourrat. *Carbon 20*:167 (1982).

20. D. Joseph and A. Oberlin. *Carbon 21*:559, 565 (1983).

21. X. Bourrat, A. Oberlin, and J. C. Escalier. *Fuel 65*:1490 (1986).

22. A. Oberlin and G. Terriere. *J. Microsc. 18*:247 (1973).

23. J. N. Rouzaud and A. Oberlin. *Carbon 27*: in press (1989).

24. A. Oberlin, G. Terriere, and J. L. Boulmier. *Tanso 80*:29 (1975). A. Oberlin, G. Terriere, and J. L. Boulmier. *Tanso 83*:153 (1975).

25. A. Oberlin, M. Endo, and T. Koyama. *J. Crystal Growth 32*:335 (1976).

26. M. Endo, A. Oberlin, and T. Koyama. *Jpn. J. Appl. Phys. 16*:1519 (1977).

27. M. Villey, A. Oberlin, and A. Combaz. *Carbon 17*:77 (1979).

28. J. L. Boulmier. Etude structurale de quelques séries de roches sédimentaires carbonées, C.N.R.S. AO 12748. Thèse d'Etat, Université d'Orléans, France, 25 juin 1976.

29. M. Villey. Simulation thermique de l'évolution des kérogènes. Thèse d'Etat, Université d'Orléans, France, 5 fevrier 1979.

30. J. L. Boulmier, A. Oberlin, J. N. Rouzaud, and M. Villey. In *Scanning Electron Microscopy*, Vol. IV (ed. SEM Inc.). AMF O'Hare, Chicago, 1982, p. 1523.

31. J. Goma and M. Oberlin. *Thin Solid Films 65*:221 (1980).

32. J. N. Rouzaud, A. Oberlin, and C. Beny-Bassez. *Thin Solid Films 105*:75 (1983).

33. A. Oberlin, J. Goma, and J. N. Rouzaud. *J. Chim. Phys. 81*:701 (1984).

34. A. Oberlin and M. Guigon. In *Fibre Reinforcement for Composite Materials,* Vol. 2 (A. R. Bunsell, ed.). Elsevier, Amsterdam, 1988, p. 149.

35. J. Goma and A. Oberlin. *Carbon 23*:85 (1985).

36. J. B. Donnet. *Carbon 20*:266 (1982).

37. J. Lahaye and G. Prado. In *Chemistry and Physics of Carbon,* Vol. 14 (P. L. Walker Jr. and P. Thrower, eds.). Marcel Dekker, New York, 1978, p. 167.

38. U. Hofmann, A. Ragoss, G. Rudorff, R. Holst, W. Ruston, A. Russ, and G. Ruess. *Z. Anorg. Chem. 255*:195 (1947).

39. C. E. Hall. *J. Appl. Phys. 19*:271 (1948).

40. H. P. Boehm. *Z. Anorg. Chem. 297*:315 (1958).

41. V. I. Kasatotchkine, V. M. Loukianovitch, N. M. Popov, and K. V. Tchmoutov. *J. Chim. Phys. 58*:822 (1961).

42. J. B. Donnet and J. C. Bouland. *Rev. Gen. Caoutchouc 41*:407 (1964).

43. J. B. Donnet and J. C. Bouland. *Carbon 4*:201 (1966).

44. J. B. Donnet, J. Schultz, and A. Eckhardt. *Carbon 6*:781 (1968).

45. F. A. Heckman and J. S. Clarke. *7th Biennial Conf. Carbon,* Cleveland (1965).

46. M. L. Rudee. *Carbon 5*:155 (1967).

47. P. A. Marsh, A. Voet, T. J. Mullens, and L. D. Price. *Carbon 9*:797 (1971).

48. A. Oberlin and G. Terriere. *J. Microsc. 14*:1 (1972).

49. G. Kaye. *Carbon 2*:413 (1965).

50. E. A. Kmetko. *Proc. First and Second Carbon Conf.,* 21 (1956), University of Buffalo, Buffalo, NY.

51. T. Tsuzuku. *Proc. Fourth Carbon Conf.,* 403 (1960), Pergamon Press, Oxford, Great Britain.

52. H. Akamatu and H. Kuroda. *Proc. Fourth Carbon Conf.,* 355 (1960), Pergamon Press, Oxford, Great Britain.

53. W. D. Schaeffer, W. R. Smith, and M. H. Polly. *Ind. Eng. Chem. 45*:1721 (1953).

54. H. T. Pinnick. *J. Chem. Phys. 20*:756 (1952).

55. M. Inagaki and T. Noda. *Bull. Chem. Soc. Jpn. 35*:1652 (1962).

56. X. Bourrat and A. Oberlin. *Carbon* (1988).

57. X. Bourrat, A. Oberlin, H. Van Damme, C. Gateau, and R. Bachelar. *Carbon 26*:100 (1988).

58. B. B. Mandelbrot. In *Fractals: Form, Chance and Dimension.* W. H. Freeman, San Francisco, 1977.

59. D. V. Badami. *Carbon 3:*53 (1965).

60. S. Iijima. *J. Solid State Chem. 42:*101 (1982).

61. A. Oberlin, M. Oberlin, and J. R. Comte-Trotet. *J. Microsc. Spectrosc. Electron 1:*391 (1976).

62. S. Bonnamy and A. Oberlin. *Carbon 20:*499 (1982).

63. J. Ayache. Simulation thermique de l'évolution des roches mères pétrolières. Pyrolyse de substances modèles. Thèse d'Etat, Université d'Orléans, France, 25 juin 1987.

64. J. Ayache, A. Oberlin, and M. Inagaki. Part I, submitted to *Carbon* (1989).

65. J. Ayache, A. Oberlin, and M. Inagaki. Part II, submitted to *Carbon* (1989).

66. F. Cortial. Les bitumes du Francevillien et leurs kérogènes. Relations avec les minéralisations uranifères. Thèse de l'Université Louis Pasteur, Strasbourg, France, 25 janvier 1985.

67. M. Audier. Composites métal-carbone obtenus par dismutation catalytique de CO. Thèse d'Etat, Université de Grenoble, France, 10 mai 1974.

68. M. Audier, A. Oberlin, M. Oberlin, M. Coulon, and L. Bonnetain. *Carbon 19:*217 (1981).

69. M. Inagaki and A. Oberlin. *Tanso 97:*66 (1979).

70. M. Inagaki, K. Kuroda, and M. Sakai. *High Temp. High Press. 13:*207 (1981).

71. M. Inagaki, K. Kuroda, and M. Sakai. *Carbon 21:*231 (1983).

72. M. Inagaki, K. Kuroda, and M. Sakai. *Carbon 22:*617 (1984).

73. P. W. Whang, F. Dachille, and P. L. Walker Jr. *High Temp. High Press. 6:*137 (1974).

74. M. Inagaki, M. I. Shihara, and S. Naka. *High Temp. High Press. 8:*279 (1976).

75. E. Fitzer, K. Mueller, and W. Schaefer. In *Chemistry and Physics of Carbon,* Vol. 7 (P. L. Walker Jr. and P. Thrower, eds.). Marcel Dekker, New York, 1971, p. 238.

76. P. G. Rouxhet, M. Villey, and A. Oberlin. *Geochim. Cosmochim. Acta 43:*1705 (1979).

77. J. Brooks and G. Shaw. *Nature 219:*532 (1968).

78. E. Fitzer, C. Holley, L. Liu, and T. Trendelenburg. *16th Biennial Conf. Carbon,* San Diego (1983), p. 38.

79. G. Bhatia, E. Fitzer, and D. Kompalik. In *Proc. Int. Conf. Carbon,* Bordeaux (1984), p. 330.

80. G. Bhatia, E. Fitzer, and D. Kompalik. *Carbon 24:*489 (1986).

81. M. Ihnatowicz, P. Chiche, J. Deduit, S. Pregermain, and R. Tournant. *Carbon 4:*41 (1966).

82. J. Millet, J. E. Millet, and A. Vivares. *J. Chim. Phys. 60:* 553 (1963).

83. J. D. Brooks and G. H. Taylor. *Carbon 3:*185 (1965).

84. J. D. Brooks and G. H. Taylor. In *Chemistry and Physics of Carbon,* Vol. 4 (P. L. Walker Jr. and P. Thrower, eds.). Marcel Dekker, New York, 1968, p. 243.

85. B. Alpern. In *Kerogen* (B. Durand, ed.). Technip, Paris, 1980, p. 339.

86. M. Teichmuller and R. Teichmuller. *Bull. Centres Rech. Explor. Prod. Elf Aquitaine 5:*491 (1981).

87. H. Marsh, F. Dachille, M. Iley, P. L. Walker, and P. W. Whang. *Fuel 52:*253 (1973).

88. BCRA Reports. Carbonization Research Reports no. 13 (1975), no. 25 (1976), no. 85 (1979), no. 82 (1980).

89. D. Auguie, M. Oberlin, A. Oberlin, and P. Hyvernat. *Carbon 19:*227 (1981).

90. S. Bonnamy. Caractérisation des produits pétroliers lors de la pyrolyse de leur fraction lourde. Thèse d'Etat, Université d'Orléans, France, 11 septembre 1987.

91. S. Bonnamy and A. Oberlin. *Carbon,* in press.

92. H. Honda, H. Kimura, and Y. Sanada. *Carbon 9:*695 (1971).

93. J. L. White. Aerospace Co. Report F 04701-73-C 0074 (1974).

94. J. L. White and J. E. Zimmer. *Carbon 16:*469 (1978).

95. M. Kleman and J. F. Sadoc. *J. Phys. Lett. 40:*L.569 (1979).

96. J. F. Sadoc and R. Mosseri. *Phil. Mag. B45:*467 (1982).

97. S. Chandrasekkar, B. K. Sadashiva, and K. A. Suresh. *Pramâna 9:*471 (1977).

98. J. Billard, J. C. Dubois, N. H. Tinh, and A. Zann. *Nouv. J. Chim. 2:*535 (1978).

99. A. Beguin, J. Billard, J. C. Dubois, and N. H. Tinh. *J. Phys. 40:*15 (1979).

100. J. Billard. In *Chemical Physics Series,* Vol. 11, Springer, Berlin, 1980, p. 383.

101. A. M. Levelut. *J. Phys. 40:*L.81 (1979).

102. C. Destrade, M. C. Mondon-Bernaud, and N. H. Tinh. *Mol. Cryst. Liq. Cryst. 49:*169 (1979).

103. N. H. Tinh, C. Destrade, and H. Gasparoux. *Phys. Lett. A72:* (1979).

104. J. M. Guet and D. Tchoubar. *Carbon 13:*273 (1985).

105. A. Oberlin, M. Oberlin, and M. Maubois. *Phil. Mag. 32:*833 (1975).

106. I. S. McLintock and J. C. Orr. In *Chemistry and Physics of Carbon,* Vol. 11 (P. L. Walker Jr. and P. Thrower, eds.). Marcel Dekker, New York, 1973, p. 243.

107. H. E. Blayden, J. Gibson, and H. L. Riley. *Proc. Conf. Ultrafine Struct. Coals and Cokes.* BCURA, London, 1944, p. 176.

108. B. E. Warren. *Phys. Rev. 59:*693 (1941).

109. R. E. Franklin. *Acta Cryst. 3:*107 (1950).

110. R. E. Franklin. *Acta Cryst. 4:*253 (1951).

111. J. Mering and J. Maire. In *Les Carbones,* Vol. 1 (Groupe Français d'Etude des Carbones, ed.). Masson, Paris, 1965, p. 129.

112. M. Shiraishi, G. Terriere, and A. Oberlin. *J. Mater. Sci. 13:*702 (1978).

113. M. Inagaki, A. Oberlin, and T. Noda. *Tanso 81:*68 (1975).

114. A. Marchand. In *Les Carbones,* Vol. 1 (Groupe Français d'Etude des Carbone, ed.), Masson, Paris, 1965, p. 140.

115. H. T. Pinnick. *Proc. 1st Carbon Conf.,* 3 (1953), University of Buffalo, Buffalo, NY.

116. A. Oberlin, J. L. Boulmier, and B. Durand. In *Advances in Organic Geochemistry.* Technip, Paris, 1973, p. 15.

117. A. Oberlin, J. L. Boulmier, and B. Durand. *Geochim. Cosmochim. Acta 38:*647 (1974).

118. J. L. Boulmier, A. Oberlin, and B. Durand. In *Proc. 7th Int. Meet. Org. Geochem., Madrid* (R. Campos, J. Goni, eds.). Enadimsa, Madrid, 1977, p. 781.

119. A. Oberlin, M. Villey, and A. Combaz. *Carbon 18:*347 (1980).

120. A. Oberlin and M. Oberlin. *J. Microsc. 132:*353 (1983).

121. A. Oberlin, S. Bonnamy, X. Bourrat, M. Monthioux, and J. N. Rouzaud. In *ACS Symposium, Series no. 303 (Petroleum Derived Carbons)* (J. D. Bacha, J. W. Newman and J. L. White, eds.). American Chemistry Society, New York, 1986, p. 85.

122. P. Chiche. *J. Chim. Phys. 60:*792 (1963).

123. S. Pregermain and P. Chiche. *J. Chim. Phys. 60:*799 (1963).

124. P. Chiche, S. Durif, and S. Pregermain. *J. Chim. Phys. 60:*825 (1963).

125. S. Durif. *J. Chim. Phys. 60:*816 (1963).

126. F. Bensaid. Corrélation entre la microtexture des charbons et des mélanges et les parametres de la cokéfaction. Thèse d'Etat, Universite d'Orléans, France, 27 juin 1983.

127. H. A. G. Chermin and D. W. Van Krevelen. *Fuel 36:*85 (1957).

128. D. Fitzgerald and D. W. Van Krevelen. *Fuel 38:*17 (1959).

129. M. Monthioux. Caractérisation de produits pétroliers lourds par traitements thermiques. Thèse 3ème Cycle, Université d'Orléans, France, 13 novembre 1980.

130. X. Bourrat, A. Oberlin, and J. C. Escalier. *Fuel 66:*542 (1987).

131. J. Gillot, B. Lux, P. Cornuault, and F. du Chaffaut. *8th Biennial Conf. Carbon,* 1967, p. 175.

132. E. Barillon. *J. Chim. Phys. 65:*428 (1968).

133. N. Christu, E. Fitzer, K. Kalka, and W. Schafer. *J. Chim. Phys. 65:*50 (April 1968).

134. E. Barillon, M. Denoux, and A. Galy. *J. Chim. Phys. 65:*162. (April 1968).

135. M. P. Whittaker and L. I. Grindstaff. *Carbon 7:*615 (1969).

136. E. Fitzer and S. Weisenburger. *Carbon 14:*195 (1976).

137. X. Bourrat, A. Oberlin, and J. C. Escalier. *C. R. Acad. Sci. Paris, 298.II.16:*787 (1984).

138. I. Mochida and H. Marsh. *17th Biennial Conf. Carbon,* 1985, p. 276.

139. I. Letizia. *High Temp. High Press. 9:*291, 297 (1977).

140. M. H. Wagner, W. Hammer, and G. Wilhelmi. *High Temp. High Press. 13:*153 (1981).

141. F. Tombrel and J. Rappeneau. In *Les Carbones,* Vol. 2 (Groupe Français d'Etude des Carbones, ed.). Masson, Paris, 1965, p. 779.

142. J. C. Bokros. In *Chemistry and Physics of Carbon,* Vol. 5 (P. L. Walker Jr., ed.). Marcel Dekker, New York, 1969, p. 1.

143. W. V. Kotlensky. In *Chemistry and Physics of Carbon,* Vol. 9 (P. L. Walker Jr. and P. Thrower, eds.). Marcel Dekker, New York, 1973, p. 173.

144. A. W. Moore. In *Chemistry and Physics of Carbon,* Vol. 11 (P. L. Walker Jr. and P. Thrower, eds.). Marcel Dekker, New York, 1973, p. 69.

145. P. A. Tessner. In *Chemistry and Physics of Carbon,* Vol. 19 (P. L. Walker Jr. and P. Thrower, eds.). Marcel Dekker, New York, 1984, p. 65.

146. P. Loll, P. Delhaes, A. Pacault, and A. Pierre. *Carbon 15:*383 (1977).

147. J. Goma and A. Oberlin. *Carbon 24:*135 (1986).

148. J. Goma. Pyrocarbones et interfaces silicium-carbone. Thèse d'Etat, Université d'Orléans, 5 mai 1983.

149. A. Pacault. In *Les Carbones,* Vol. 1 (Groupe Français d'Etude des Carbones, ed.). Masson, Paris, 1965, p. 479.

150. D. B. Fischbach. In *Chemistry and Physics of Carbon,* Vol. 7 (P. L. Walker Jr., ed.). Marcel Dekker, New York, 1971, p. 1.

151. A. Pacault. In *Chemistry and Physics of Carbon,* Vol. 7 (P. L. Walker Jr., ed.). Marcel Dekker, New York, 1971, p. 107.

152. J. Abrahamson. *Carbon 11:*337 (1973).

153. J. Abrahamson. *Carbon 12:*111 (1974).

154. A. Oberlin and G. Terriere. *Carbon 13:*367 (1975)L

155. S. de Fonton. Croissance cristalline des carbones sous pression. Thèse de Docteur-Ingénieur, Université d'Orléans, France, 5 juillet 1978.

156. M. Inagaki, A. Oberlin, and S. de Fonton. *High Temp. High Press. 9:*453 (1977).

157. S. de Fonton, A. Oberlin, and M. Inagaki. *J. Mater. Sci. 15:* 909 (1980).

158. M. Bonijoly, M. Oberlin, and A. Oberlin. *Int. J. Coal Geol. 1:*283 (1982).

159. A. Deubergue, A. Oberlin, J. H. Oh, and J. N. Rouzaud. *Int. J. Coal Geol. 8:*375 (1987).

160. J. H. Oh. Etude structurale de la graphitation naturelle. Thèse de l'Université d'Orléans, France, 23 octobre 1987.

161. M. Inagaki and S. Naka. *J. Mater. Sci. 10:*814 (1975).

162. R. T. K. Baker and P. S. Harris. In *Chemistry and Physics of Carbon,* Vol. 14 (P. L. Walker and P. Thrower, eds.). Marcel Dekker, New York, 1978, p. 83.

163. M. Endo. Mécanisme de croissance en phase vapeur de fibres de carbone. Thèse de l'Université d'Orléans, France, 9 octobre 1975.

164. T. Koyama, M. Endo, and Y. Omima. *Jpn. J. Appl. Phys. 11:*445 (1972).

165. M. Audier, M. Coulon, and A. Oberlin. *Carbon 18:*73 (1980).

166. M. Audier, A. Oberlin, and M. Coulon. *J. Cryst. Growth 55:*549 (1981).

167. M. Audier, A. Oberlin, and M. Coulon. *J. Cryst. Growth 57:*524 (1982).

168. G. G. Tibbets. *J. Cryst. Growth 66:*632 (1984).

169. S. D. Robertson. *Carbon 8:*365 (1970).

170. R. T. K. Baker, P. S. Harris, F. S. Feates, and R. J. Waite. *J. Catal. 26:*51 (1972).

171. E. L. Evans, J. M. Thomas, P. A. Thrower, and P. L. Walker. *Carbon 11*:441 (1973).

172. H. P. Boehm. *Carbon 11*:583 (1973).

173. T. Baird, J. R. Fryer, and B. Grant. *Carbon 12*:591 (1974).

174. D. J. Johnson. In *Chemistry and Physics of Carbon*, Vol. 20 (P. L. Walker Jr. and P. Thrower, eds.). Marcel Dekker, New York, 1987, p. 1.

175. A. Oberlin and M. Guigon. In *Science and New Applications of Carbon Fibers*, November 19, 1984, Toyohashi University of Technology (Japan), p. 1.

176. M. Guigon, A. Oberlin, and G. Desarmot. *Fibre Sci. Technol. 20*:55 (1984).

177. M. Guigon, A. Oberlin, and G. Desarmot. *Fibre Sci. Technol. 20*:177 (1984).

178. M. Guigon and A. Oberlin. *Composites Sci. Technol. 25*:231 (1986).

179. M. Guigon and A. Oberlin. *Composites Sci. Technol. 27*:1 (1986).

180. A. Oberlin and M. Oberlin. *Revue Chim. Miner. 18*:442 (1981).

181. J. B. Donnet and R. Bansal. *Carbon Fibers,* Vol. 3 (Menachim Lewin, ed.). Marcel Dekker, New York, 1984.

182. G. E. Bacon. *Acta Crystallogr. 4*:558 (1951).

183. G. W. Brindley and J. Mering. *Acta Crystallogr. 4*:441 (1951).

184. X. Bourrat, A. Oberlin, and J. C. Escalier. *Fuel 66*:542 (1987).

2

Mechanisms and Physical Properties of Carbon Catalysts for Flue Gas Cleaning

HARALD JÜNTGEN

*Bergbau-Forschung GmbH[*] and University of Essen,
Essen, Federal Republic of Germany*

HELMUT KÜHL

Bergbau-Forschung GmbH, Essen, Federal Republic of Germany

---

[*]Retired.

I. INDUSTRIAL SIGNIFICANCE AND STATE OF RESEARCH

Activated carbon can be used as a catalyst support [1] and as a catalyst. Examples of reactions in which the carbon itself acts as catalyst are shown in Fig. 1. The chlorination and dehydrochlorination reactions [2] are performed in the chemical industries to produce important chemical intermediates such as phosgene [3], sulfuryl chloride [4], and chlorinated olefins and paraffins. Unfortunately, very little scientific information on the mechanism of reactions and on the catalytic activity for these reactions as a function of the properties of the carbon catalyst is available.

More information is known on the role of carbon catalysts in oxidation reactions. The oxidation of H_2S to sulfur is available for the removal of H_2S from oxygen-containing gases. The oxidation product, sulfur, can be recovered by extraction with organic solvents or by thermal treatment of the sulfur-loaded activated carbon. The reaction mechanisms of both catalytic oxidation and desorption of the oxidation product sulfur have been discussed in several papers [5,6].

OXIDATION IN GAS PHASE   CHLORINATION

$$SO_2 \xrightarrow{H_2O, O_2} H_2SO_4 \qquad CO + \xrightarrow{Cl_2} COCl_2$$

$$NO \xrightarrow{O_2} NO_2 \qquad SO_2 + \xrightarrow{Cl_2} SO_2Cl_2$$

$$NO \xrightarrow{O_2, NH_3} N_2, H_2O \qquad HC \equiv CH \xrightarrow{Cl_2} CHCl = CHCl$$

$$H_2S \xrightarrow{O_2} S, H_2O$$

DEHYDROCHLORINATION

OXIDATION IN LIQUID PHASE

$$CHCl_2-CHCl_2 \xrightarrow{-HCl} CHCl=CCl_2$$

$$H\ SO_3^- \xrightarrow{O_2} SO_4^{--}$$

FIG. 1 Typical oxidation and chlorination reactions catalyzed by carbon catalysts.

The oxidation of sulfur dioxide to sulfuric acid on active coke is the basic reaction for the dry SO_2 removal from flue gases. The oxidation of NO to NO_2 is the starting reaction for the catalytic conversion of NO with ammonia to nitrogen and water vapor, which is the basic reaction of NO_x removal from flue gases. Here activated carbon is in a position to catalyze this important reaction in the temperature range of 100-150°C in competition with SCR[*] catalysts (V_2O_5 on TiO_2 supports), which perform the reaction at temperatures higher than 330°C. These reactions with active coke or activated carbon are industrially applied as processes for the simultaneous removal of SO_2 and NO_x, and for the removal of nitrogen oxides from flue gases downstream of an SO_2 scrubbing process in Japan and West Germany [7].

Based on intensive, continuous, fundamental research, large parts of the mechanisms and kinetics of these reactions are known. Further, there is some experimental knowledge of the catalytic activity as a function of catalyst properties and reaction conditions. Therefore this chapter deals with the first three reactions in Fig. 1 and their interaction under the conditions of flue gas cleaning. After presentation of the manufacture and general properties of carbon catalysts, the reactions of the system $NO-O_2-H_2O-NH_3-C$ are discussed. Then the system $SO_2-O_2-H_2O-C$ is described, including adsorption and regeneration cycles. Finally, the measurements of interaction of both systems are presented.

II. CARBON AS CATALYST FOR OXIDATIVE REACTIONS

A. General Features

A general precondition for application of a carbon catalyst is that the carbon is inert and stable against the reactants during the performance of the catalytic reaction. For reactions in which oxidative steps are included, the stability of the carbon catalyst against oxygen is of great significance. Therefore special tests must be performed in order to investigate the reaction behavior of carbon with oxygen. The reaction rate of investigated carbons against

---

[*]Selective catalytic reductions.

oxygen is influenced by many parameters, such as, on the one hand, the oxygen concentration and the temperature and, on the other hand, the properties of the carbon catalysts used. A typical test for the characterization of the reactivity of carbon against oxygen is the determination of its ignition temperature under standard conditions. It has been observed that the temperature of ignition is dependent on the precursor material and the production conditions. For active coke the ignition temperature increases with the temperature of carbonization. Active coke and activated carbons made from bituminous coal are generally stable against oxygen below about 200°C.

Further, the preconditions of practical use of catalysts are favorable mechanical properties such as strength and low flow resistance. In general, catalysts formed by an extrusion process have less flow resistance and a higher mechanical strength than broken-grain fractions. All these properties can be attained with catalysts manufactured according to the process described in Section IIB. The mechanical strength is very important, because the modern flue gas cleaning processes work in moving beds for the adsorption and desorption step, such that the attrition loss cannot be used again in the process. Therefore the economy of the process is strongly negatively correlated with the loss of carbon catalyst by attrition.

The relationship between the catalytic activity of carbon catalysts—e.g., activated carbons or active coke—and their properties is complex and is influenced by the special features of the chemical reactions considered. However, some general facts will be discussed here. A significant factor influencing catalytic properties of activated carbon is surely the degree of order of the graphitic lattice. Activated carbons have only small areas of disordered graphitic structure elements. Their catalytic activity is characterized by electron transfer from the edges of the carbon sheets. Therefore the presence and distribution of these active sites is important. The catalytic activity of graphitic structure can also be changed by the presence of heteroatoms such as sulfur and nitrogen in the graphitic lattice. Nitrogen mostly prohibits activity [8-10], while sulfur inhibits activity for oxidation. The catalytic activity can also be influenced by the presence of acidic or basic surface oxide complexes. Also, impurities from mineral compounds can act as catalyst poisons or as promotors of catalytic activity. Furthermore, the pore structure has significance for adjustment of catalytic activity of activated carbon.

B. Manufacture and Pore Structure of Carbon Catalysts

There are many industrial processes for the production of activated carbon from different precursors such as wood, peat, lignite, bituminous coal, coconut shells, etc. [11,12]. It is known that pellets made from preoxidized bituminous coal are excellent activated carbons and catalysts.

A simplified flow sheet of the production of activated carbon made from bituminous coal by steam activation is given in Fig. 2 [13,14]. The finely ground coal is oxidized by air at < 300°C, extruded with a binder, carbonized, and activated with steam. This gives cylindrical carbon pellets with the diameters of 1-10 mm. The degree of activation (burn-off) can be varied within wide limits. A washing process with mineral acids can be used to remove impurities. The carbonization product is called active coke; the activated materials are called activated carbons.

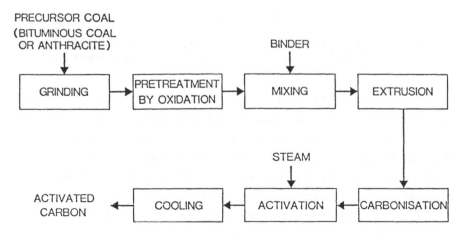

FIG. 2 Production steps of formed active coke and activated carbon made from bituminous coal by steam activation.

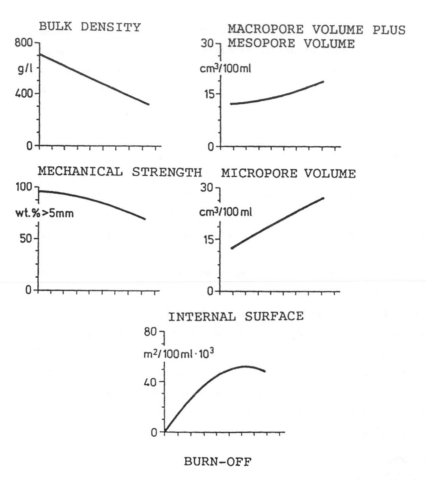

FIG. 3 Mechanical properties, pore volume, and internal surface relative to burn-off for activated carbons made from bituminous coal.

Figure 3 gives the most important properties of activated carbons as a function of burn-off, such as bulk density, mechanical strength, volume of macro- and micropores, and internal surface. It is evident that the micropore volume per 100 ml bulk volume is strongly influenced by the burn-off and increases linearly with burn-off, whereas the macropore plus mesopore volume is changed neglibly. In more detail, it has been found that the mesopore volume increases only slightly with the burn-off, whereas the macropore volume is nearly

FIG. 4 Scanning electron micrograph of carbonized coal particles
that had been previously oxidized by air at temperatures between
240 and 300°C. Carbonization temperature: 900°C.

constant within experimental error. The internal surface per 100 ml
goes through a maximum. These figures show that pore volume, pore
size distribution, and internal surface can be widely varied by the
burn-off. Also, the precursor material and the production process
[15] can influence the properties of activated carbon.

The properties of active coke and activated carbons can also be
confirmed by scanning electron micrographs, which are shown in Figs.
4-7. Figure 4 shows particles of ground, oxidized and carbonized
bituminous coal under the scanning electron microscope. It is characteristic that no swelling occurs, that no gasification pores are
formed by volatilization, and that the original shape of the oxidized
coal particles is not changed during carbonization, provided that the
degree of oxidation is high enough. Particles of carbonized tar pitch,
which is used as binder, are shown in Fig. 5. The carbonization leads
to the well-known structure of pitch coke. Figures 6 and 7 show

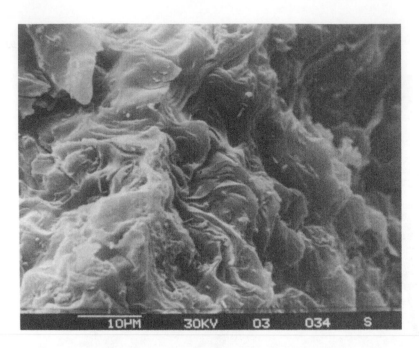

FIG. 5 Scanning electron micrograph of carbonized pitch.
Carbonization temperature: 900°C.

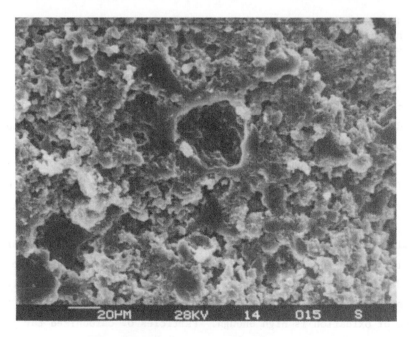

FIG. 6 Scanning electron micrograph of a section of an active
coke pellet after carbonization.

FIG. 7 Scanning electron micrograph of a section of an activated carbon pellet after carbonization and steam activation.

sections of pellets after carbonization and after activation, respectively. The pellets consist of compacted particles of char formed from the oxidized coal and held together by coke residue from the tar binder. Differences between the carbonized and the steam-activated product cannot be detected by the scanning electron microscope. This confirms that the macropore system is scarcely changed by the reaction of the pellets with steam, as shown in Fig. 3. The reaction of carbon with steam takes place within the char particles, creating new meso-, micro-, and submicropores and expanding the network of micropores already developed during carbonization.

These observations lead to the schematic presentation of the catalyst pellets shown in Fig. 8. They consist of char particles formed by the carbonization of oxicoal and held together by pitch coke bridges. The steam activation step forms meso-, micro-, and submicropores in the char particles and has little or no effect on

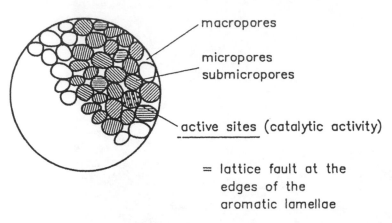

FIG. 8 Schematic presentation of the carbon catalyst pellet.

the size of the macropores between char particles and pitch coke
bridges. Active sites responsible for catalytic reaction may be
distributed in the pores within the char particles. In general the
active surface area (ASA) (see Section IVC) for catalytic reactions
is much lower than the total surface area (TSA) of the pore system
formed in the char particles.

   The process of creating efficient carbon catalysts consists of
manufacturing carbon pellets with a network of macro- and mesopores
that allows reactants to enter rapidly in the micro- and submicropores of char particles on the one hand, and of creating active sites
in the char pellets that are selective for the chemical reaction on
the other hand. In the next sections the mechanisms and kinetics of
different chemical reactions will be discussed and the correlations
between the activity for this reaction and other catalyst properties
will be shown.

III. MECHANISM AND KINETICS OF NO CONVERSION WITH AMMONIA

The overall reaction

$$4NO + 4NH_3 + O_2 \longrightarrow 6H_2O + 4N_2 \qquad\qquad (1)$$

is suitable for the removal of NO from flue gases. It is mostly catalyzed with TiO_2-supported metals at temperatures up to 350°C. In many cases it is more advantageous to perform the reaction of Eq. (1) at lower temperatures, say 100°C, to use this removal system downstream of a wet SO_2 removal. It has also been found that carbon catalyzes the reaction of Eq. (1), especially at low temperatures, between 100 and 150°C. This reaction is also useful as a model reaction to study the relation between reaction rate and the parameters of manufacture and on catalyst properties.

Two difficulties occur. First, the overall reaction is complex and consist of several separate steps running in parallel or consecutive reactions. From this, it follows that products other than nitrogen and water mentioned in Eq. (1), such as HNO_3, can be formed. Second, side reactions between the catalyst surface and gaseous components such as NO, O_2, H_2O, and NH_3 take place, forming or destroying intermediate surface oxide complexes, which change the surface properties of the catalyst. These reactions depend on the reaction conditions, especially on temperature and the concentrations of oxygen and water vapor. Further, they are different for the types of catalyst, e.g., activated carbon or active coke. These problems will both be discussed next in more detail.

A. Reaction Mechanism of Reduction

It has been found that the first step of the reduction of NO by ammonia is the oxidation of NO to NO_2 at the catalyst surface. The reaction rate of this starting step can be influenced by the amount and kind of surface oxides. The consecutive reactions depend strongly on the reaction temperature. At low temperatures part of the NO_2 adsorbed is converted into ammonium nitrate, which is enriched on the carbon surface. In consequence, only part of the NO can be converted into nitrogen and water via ammonium nitrite. In contrast, at temperatures above 100°C all NO_2 adsorbed reacts to form N_2 and H_2O. As a result, the catalytic reaction can only be performed in commercial units at temperatures above 100°C. The mechanism identified so far is shown in Table 1 [16]. Latest results show that it depends also

TABLE 1   Postulated Mechanisms for the NO Reaction
with Ammonia at Different Temperatures

$$2NO_{gas} \rightarrow 2NO_{ads} \qquad (1)$$

$$2(NO_{ads} + \tfrac{1}{2}O_2 \rightarrow NO_{2\ ads}) \qquad (2)$$

Preferential temperatures < 50°C

$$2NO_{2\ ads} \rightarrow N_2O_{4\ ads} \qquad (3)$$

$$N_2O_{4\ ads} + 2NH_3 + H_2O \rightarrow NH_4NO_3 + NH_4NO_2 \qquad (4)$$

$$NH_4NO_2 \rightarrow N_2 + 2H_2O \qquad (5)$$

$$2NO + O_2 + 2NH_3 \rightarrow NH_4NO_3 + N_2 + H_2O \qquad (6)$$

Preferential temperatures > 100°C

$$2(NO_{2\ ads} + NO \rightarrow N_2O_{3\ ads}) \qquad (7)$$

$$2(N_2O_{3\ ads} + H_2O + 2NH_3 \rightarrow 2NH_4NO_{2\ ads}) \qquad (8)$$

$$4(NH_4NO_2 \rightarrow N_2 + 2H_2O) \qquad (5)$$

$$4NO + O_2 + 4NH_3 \rightarrow 4N_2 + 6H_2O \qquad (9)$$

on the pH value at the carbon surface.   High pH values favor the
formation of nitrate.   Low pH values favor the reaction to nitrogen
and water at lower temperatures down to 70°C.   The overall activation
energy was found to be 0-7 kJ mol^{-1} for different carbon catalysts.
This low value indicates that the first step—the adsorption of NO
and the oxidation to NO_2—is strongly influenced by the adsorption
step, which has a negative temperature dependance.   This mechanism
is indicated by the following experimental results:

1. NO_2 is identified as an intermediate that is strongly
     adsorbed at the carbon surface as a function of temperature.

2.   The conversion to N_2 and H_2O takes place preferentially in
     the presence of oxygen.

3.   The stoichiometric NH_3/NO ratio was found to be nearly equal
     to 1 using activated carbons as catalysts.

4.   Formation of NO_3^- is confirmed only at low temperatures.

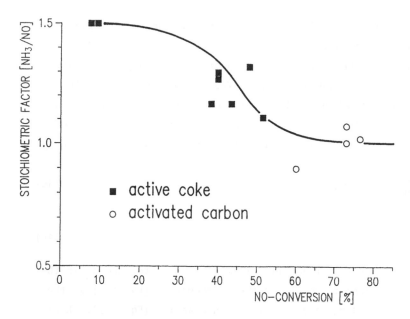

FIG. 9 NH_3/NO stoichiometric factor as a function of reactivity for active coke and activated carbon.

In the case of active coke the NH_3/NO ratio can increase up to 1.5 (see Fig. 9), which can be explained by side reactions between ammonia and surface oxides (see Section VI).

B. Kinetics and Interaction of Main Reaction [Eq. (1)] with Side Reactions at the Carbon Surface

The following side reactions are possible and are performed depending on reaction conditions and the type of carbon catalyst.

$$2NO + C \rightarrow N_2 + CO_2$$
$$2NO + 2C \rightarrow N_2 + 2CO \tag{2}$$

$$2NH_3 + 3/2O_2 \rightarrow N_2 + 3H_2O \tag{3}$$

Equation (2) leads, in combination with Eq. (1), to a lower stoichiometric ratio; in contrast, the other side reaction, Eq. (3), leads to an increase of the NH_3/NO ratio and was only found with active coke as catalyst. In the following we will discuss the interaction between Eq. (1) and Eq. (2) at an activated carbon catalyst.

Recently Kleinschmidt [17] studied the mechanism and kinetics of these two reactions, using a special activated carbon made by Bergwerksverband (BWV) as a catalyst over a broad range of temperature and O_2 and H_2O concentrations. The NH_3/NO ratio in the inlet feedstream had also been varied over a wide range. His experiments and calculations lead to the following system of rate expressions. In a fixed-bed reactor the balance of educts NO and NH_3 is given by

$$\frac{V}{p_s A} \frac{dc_{NO}}{dz} = -(r_1 + r_2) \tag{4a}$$

$$\frac{V}{p_s A} \frac{dc_{NH_3}}{dz} = -r_2 \tag{4b}$$

where A is the area of reactor, V is the volume rate, and p_s is the bulk density. Here r_1 refers to the NO reduction [Eq. (2)] at the carbon surface and r_2 refers to the main reaction [Eq. (1)]. Kleinschmidt has explained the role of surface oxides for both reactions, which cannot be discussed here in detail and is described elsewhere [17]. The formal rate expression of r_1 and r_2 fits the experimental data very well, as shown in Figs. 10-12.

$$r_1 = \frac{k_1 c_{NO}^2 c_{O_2} c_{H_2O}}{(1 + k_2 c_{NO}^2 c_{O_2} + k_3 c_{NO}^2)} \tag{5}$$

$$r_2 = \frac{k_4 c_{NO}^2 c_{O_2} c_{NH_3}^{0.1}}{(1 + k_5 c_{NO}^2 c_{O_2} + k_6 c_{H_2O})^2} \tag{6}$$

Figure 10 shows the relative concentration profiles of NO in the fixed-bed reactor with a diameter of 5 cm, taking the NH_3/NO ratio (β) as the parameter. It can be seen that for β = 0 the NO conversion is very small, which means the reaction of Eq. (2) has little influence on the total reaction [Eq. (1) and Eq. (2)] of the NO reduction. By comparison with the experimental points for β = 1 and β = 2, it can be seen that both conversions as a function of residence time are

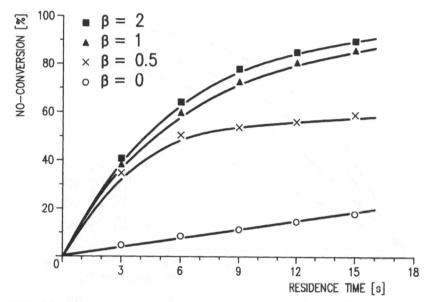

FIG. 10 Influence of inlet NH$_3$/NO feed concentration on the NO conversion at different residence times (starting concentrations, NO, 420 vpm; H$_2$O, 10% by volume; O$_2$, 6% by volume; 100°C) (Kleinschmidt).

nearly identical. That means that an excess of ammonia has no effect on the conversion or, in other words, the reaction order of NH$_3$ is nearly zero, as was also shown in the rate expression of r$_2$.

The influence of oxygen and water vapor on the conversion is shown in Figs. 11 and 12. As mentioned before, the oxygen content strongly enhances the conversion of NO. But it can be seen, too, that the effect is mainly restricted to the concentration range between 1% and 6% by volume. At higher oxygen concentrations there is no more influence. The concentration of water vapor inhibits the conversion of NO mainly in the range between 0% and 10% by volume.

For comparison of reaction rates with catalyst properties (Section IVB) the concentrations of water vapor and oxygen and the space velocity were kept constant. Thus the conversion is a parameter that describes the carbon reactivity for the main reaction [Eq. (1)]. In Section IVC a very simple rate expression (reaction order related to NO = 1 and related to NH$_3$ = 0) is used to determine a reaction

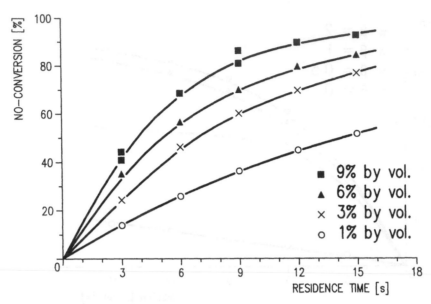

FIG. 11   NO conversion as a function of residence time at different
oxygen concentrations (starting concentrations, NO, 430 vpm; NH$_3$,
460 vpm; H$_2$O, 10% by volume; 100°C) (Kleinschmidt).

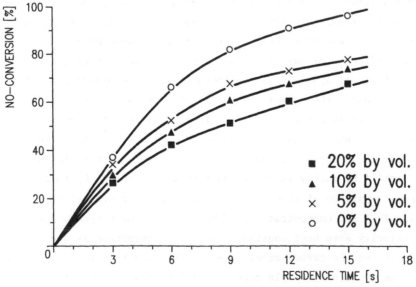

FIG. 12   NO conversion as a function of residence time at different
water vapor concentrations (starting concentrations, NO, 460 vpm;
NH$_3$, 480 vpm; O$_2$, 6% by volume; 95°C) (Kleinschmidt).

rate constant in a simple way. It was proved by varying the space velocity that was allowed under this condition, especially excess of ammonia.

Another reaction model for active coke as catalyst is used in Section VI to describe the influence of the reaction between sulfuric acid and ammonia on the system $NO-NH_3$ with constant H_2O and O_2 concentrations. Here the side reaction with NH_3 and carbon is also considered, which was only observed at active cokes (see also Fig. 9).

IV. CORRELATION BETWEEN PROPERTIES OF THE CARBON CATALYST AND THE REACTIVITY OF THE NO/NH_3 REACTION

Carbonization products and activated carbons produced by Bergwerks-verband (BWV) as described above were used as samples. The following investigations were performed:

1. Determination of pore structure by mercury porosimetry and measurement of the BET surface according to DIN 66132.
2. Determination of the activity of catalysts for NO_x reduction by NH_3 in a fixed-bed reactor in presence of oxygen and water vapor (for conditions see Table 2).
3. Measurement of reactivity of carbon with steam at elevated temperatures [18].
4. Determination of the total reflectance distribution and the mean random reflectance by an image analyzer (TAS, Leitz).
5. Determination of active surface area.

TABLE 2 Conditions for the Experiments on Catalytic NO Reduction with Ammonia in Figs. 13, 14, 18, and 19

Gas composition	
Nitrogen oxide (NO)	425 vpm
Ammonia	525 vpm
Oxygen	6.0% by volume
Water vapor	10.0% by volume
Nitrogen	Remainder
Temperature	100°C
Space velocity	500 h^{-1}

A.   Correlation between the Reactivity of the Catalyst
     and Its Manufacture

1. *Influence of Heat Treatment Temperature (HTT)*

Catalytic activity of carbonaceous catalysts depends on heat treatment during their production (carbonization step). Typical results
of measurements of NO conversion for active cokes at different carbonization temperatures are summarized in Fig. 13. This figure also
shows the reflectance distribution of the three samples. The original
carbonization product has a high catalytic activity (72% NO conversion), a mean random reflection R_r = 4.4%, and a nearly symmetrical
reflectance distribution. Thermal treatment at 800°C and 900°C
increases the R_r value to 6.1% and 6.3%, respectively, and the distribution becomes distinctly asymmetrical. In the same sequence the
catalytic activity decreases significantly, corresponding to NO conversions of 42% and 29%, respectively.

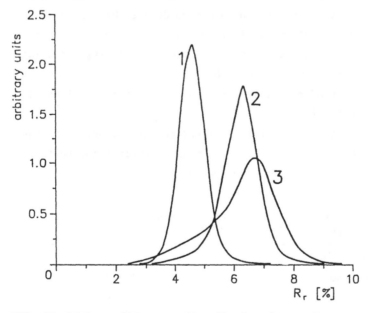

FIG. 13 Light reflectance distribution for active cokes that have
been carbonized at different temperatures [T(1), T(2), T(3)]. Mean
random reflectance, R_r (%); NO conversion (for experimental conditions see Table 2), C_{NO} (%). Sample 1: R_r 4.4%, C_{NO} 72%. Sample 2:
R_r 6.1%, C_{NO} 42%. Sample 3: R_r 6.3%, C_{NO} 29%.

2. Influence of Burn-Off

The results obtained for the samples that were investigated in this set are presented in Fig. 14 and their physical properties are listed in Table 3. Another active coke, which has been produced at a higher than optimum carbonization temperature, was used as precursor material for activation. The values of carbon burn-off lie between 0% and 20% for samples 1-6. The R_r values decrease with increasing burn-off, and the shape of the distribution changes in a characteristic manner. The gradient becomes steeper and distinct tailing occurs. The catalytic activity initially increases with increasing burn-off, then passes through a maximum and finally decreases sharply. The volume of macropores (pore size greater than 10 nm) measured by mercury porosimetry and related to the bulk volume is independent of the

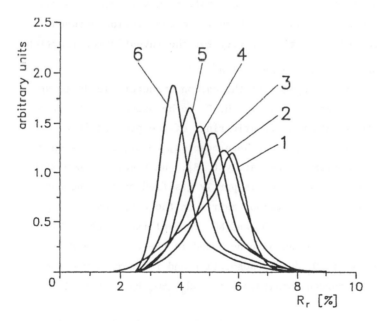

FIG. 14 Light reflectance distribution of (1) active coke and (2-6) activated with increasing burn-off from 2 to 6. Mean random reflectance R_r (%); NO conversion (for experimental conditions see Table 2) C_{NO} (%). Sample 1: R_r 5.2%, C_{NO} 62%. Sample 2: R_r 5.4%, C_{NO} 73%. Sample 3: R_r 5.1%, C_{NO} 70%. Sample 4: R_r 4.9%, C_{NO} 70%. Sample 5: R_r 4.6%, C_{NO} 68%. Sample 6: R_r 4.1%, C_{NO} 64%.

TABLE 3 Specific Macropore Volume and BET Surface Area of
(1) Active Coke and (2-6) Activated Carbons, with Increasing
Burn-Off from Sample 2 to Sample 6

| Sample | Specific macropore volume (cm^3/l) | BET surface area | |
		m^2/g	m^2/cm^2
1	180	25	16
2	176	400	241
3	180	455	263
4	180	575	322
5	172	620	332
6	181	720	371

burn-off. The BET surface (DIN 66132) rises with increasing burn-off when related to sample weight and catalyst bulk volume. As can be seen by the values in Table 3, there is no correlation between internal surface and catalytic activity for the investigated reaction.

3. Influence of Nitrogen Incorporation

The incorporation of nitrogen into the carbon structure leads to an increase of the catalytic activity. H. P. Boehm assumes that nitrogen directly enters into the aromatic rings of the pregraphitic structure by forming N-containing heterocycles, resulting in a change of electronic properties of the catalyst [8]. He treats carbon with NH_3 at 900°C, whereby the carbon is partly gasified on the one hand and nitrogen is bound in the carbon matrix on the other hand. Experiences with the simultaneous SO_2 and NO_x removal (see Section VI) [19] reveal that nitrogen can also be bound in the carbon matrix in the presence of $(NH_4)_2SO_4$ during adsorption and regeneration cycles or by treatment of carbon catalysts with $(NH_4)_2SO_4$ by a special method.

A correlation between the catalytic NO conversion by ammonia and the N content can be seen from Fig. 15. Here the conversion of NO with ammonia on an activated carbon catalyst containing different amounts of N is plotted against the N content. It can be seen that

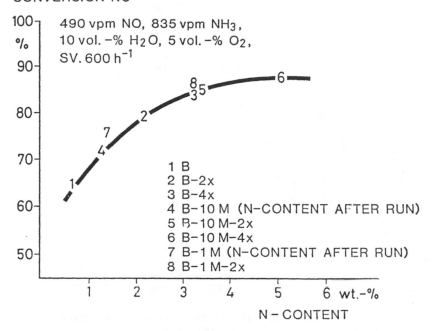

CONVERSION NO

490 vpm NO, 835 vpm NH₃,
10 vol.-% H₂O, 5 vol.-% O₂,
SV. 600 h⁻¹

1 B
2 B-2x
3 B-4x
4 B-10 M (N-CONTENT AFTER RUN)
5 B-10 M-2x
6 B-10 M-4x
7 B-1 M (N-CONTENT AFTER RUN)
8 B-1 M-2x

FIG. 15 NO conversion by NH₃ at 100°C for different activated
carbons as a function of the N content [19].

the catalytic activity is increasing nearly linearly with the N
content in the range between 1% and 3%. Higher values of N content
lead to a smaller increase of activity.

B. Significance of Reflectance

The increase of reflectance, and thus higher aromaticity values of
the char, with rising HTT can be interpreted in terms of the Oberlin-
Rouzoud concept [20]. According to their investigations, basic
structural units (BSU) are formed during the first stage of high
temperature treatment. The BSU consist of polycyclic aromatics with
5-11 rings. The units consist of single planes with maximum dimensions of about 1 nm or of stacks with two or three planes on top of
one another. They are randomly oriented at first and in the stage
of semicoke—which is the most interesting for the problem treated
here—they begin to orient in parallel as shown in Fig. 16. This

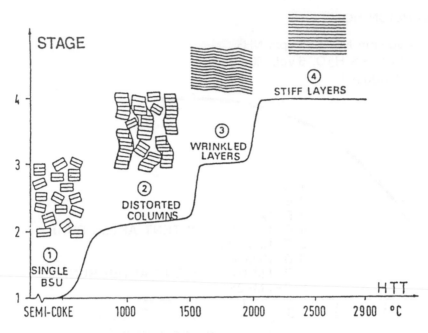

FIG. 16 Structural stages of carbonization and graphitization, drawn from 002 lattice-fringe images (Oberlin).

process of local molecular orientation (LMO) depends on HTT and on the chemical composition of the precursor material, especially on its oxygen-containing groups. The latter can be widely varied by the conditions of preoxidation during production.

Now the reasons for the catalytic activity decrease with increasing reflectance will be discussed. It cannot be assumed that each C atom, located at the internal surface of the char particles within the catalyst pellet, is active, but the active sites exist at the edges of the polynuclear ring systems. If the basic structure units are smaller and have a lower local molecular orientation, more active sites exist per volume of the char particles. Thus, the latter show a higher catalytic activity. The size of the BSU and their LMO increases while the number of active sites decreases if the heat treatment is carried out at higher temperatures. This explains the decrease of catalytic activity due to increasing HTT.

The second question is how the char particles behave during steam activation. Surprisingly, the conversion of NO increases first, passes through a maximum, and then decreases significantly during higher burn-off values by the steam/carbon reaction (Fig. 14).

Steam activation has a similar effect as HTT, and with a longer treatment (higher burn-off values) greater BSU and higher LMO occur. This could explain the decrease of active sites during higher burn-off. Another possible explanation could be that in the range of higher burn-offs the reflectance is not only influenced by the size of the BSU and their LMO, but also by the microporosity of the activated carbons, which increases with rising burn-off. In this way the reflecance decreases (diffuse light reflection) with increasing microporosity, and this effect is more sensitive than the influence of the BSU and LMO.

It was shown in Fig. 14 that the reflectance distributions of samples with higher burn-off have a less distinct symmetry than those with lower burn-off. The steep side shifts to lower reflection ranges with increasing burn-off. This behavior is attributed to the different activity of the preoxidized coal char particles and the coke of the binder.

It can be seen in Fig. 17 that indeed the reactivity of tar pitch coke toward steam is very much lower than that of the whole active coke pellet consisting of char particles and bridges between them. The reactivities were measured at 4×10^6 N/m^2 in a steam atmosphere. It can be assumed that the observed relationship holds for low-pressure steam activation during the industrial production of activated carbon. Due to the lower reactivity, the pitch coke is enriched within the pellets by burn-off. Thus, the shoulders of the reflectance distribution develop.

Figure 18 [21] summarizes all measured values of catalytic activity as a function of the mean random reflectance value R_r. It is surprising that all points can be fitted by one curve, considering the experimental error. Catalysts in the range of mean random reflecance values between 4% and 5.5% show a maximum of catalytic reactivity independent of the production including only carbonization or addi-

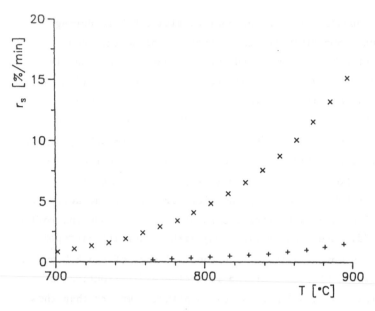

FIG. 17 Reactivity of active coke and pitch coke to water vapor
during time-linear heating (Mühlen). Conditions: 40 bar; heating
rate 10 K/min. x, Carbon catalyst; +, pitch coke.

tional steam activation, and independent of the internal surfaces
observed. This behavior can only be explained by the fact that
catalysts with the highest concentration of active sites, formed
by an optimum carbonization or activation, show R_r values in the
range mentioned.

It is very interesting to see that the carbon catalysts doped
with N do not fit the curve of reactivity versus mean random reflectance, but show a higher catalytic reactivity compared with samples
without N incorporation. Possibly the effect of change of electronic
properties by the incorporation of N in the heterocycles of pregraphitic structure is only reflected by chemical properties but not
by random reflectance.

C. Importance of Active Surface Area

To investigate the influence of the active surface area (ASA), the
catalytic activity of different carbon catalysts relative to NO
reduction with ammonia in the presence of oxygen and water vapor

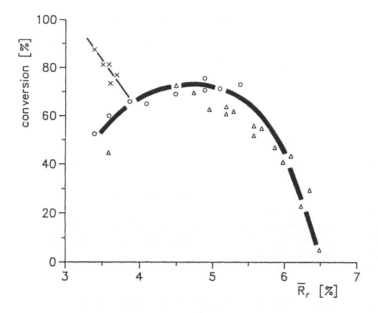

FIG. 18 NO conversion on different carbon catalysts as a function of mean random reflectance R_r. ($\triangle$) Active coke; (o) activated carbon; (x) activated carbons with incorporated nitrogen (conditions, see Table 2).

were characterized by total surface area (TSA) and active surface area. The carbons used were prepared from an oxidized bituminous coal as shown in Section IIB. Carbonization took place at temperatures between 700 and 900°C. There were different degrees of burn-off, and one sample (E) had a very high content of incorporated nitrogen (5 wt.%). The method of preparation of this sample is described in another paper [22].

The TSA values of the samples were measured by the BET method using N_2 adsorption at 77 K. The method for determining the ASA was described previously [23]. After cleaning the carbon surfaces at 950°C *in vacuo* for 3 h to remove essentially all of the oxygen surface complexes that were present as result of manufacture, the ASA values of the samples were determined from the amount of oxygen chemisorbed in 24 h at 300°C using an initial O_2 pressure of 50 Pa. The assumption was made that one oxygen atom per edge carbon atom

TABLE 4   Physical and Catalytic Properties of Different Carbon Catalysts

Samples	T_{max}[a] (K)	C_{NO}[b] (%)	k (h^{-1})	TSA (m^2/g)	ASA[c] (m^2/g)
A	973	70	722	20	6
B	1173	29	205	15	10.7
C	1173	50	416	1400	20.5
D	1173	65	630	700	40.4
E	1173	87	1224	900	52.8

[a]Maximum carbonization temperature.
[b]Conversion of NO. Experimental conditions: NO, 425 vpm; NH_3, 525 vpm; O_2, 6% by volume; H_2O, 10% by volume; rest N_2. Temperature 473 K; space velocity 600 h^{-1}.
[c]After treatment at 1123 K in vacuum for 3 h.

is chemisorbed. Further, it was assumed that the edge carbon atom occupies an area of 8.3 x 10^{-2} nm^2.

The catalytic activity was measured in a laboratory-scale screening apparatus as described in Section III. Table 4 lists the maximum temperatures T_{max} that the samples have seen during carbonization or steam activation, the conversion rates of NO at constant conditions and the calculated rate constants k. Table 4 shows that there is no correlation between the TSA and the catalytic activity, indicated by the rate constant. Considering samples B, C, D, and E, a linear relationship between the rate constant of the catalytic NO reduction with ammonia and the ASA is observed (Fig. 19a), and for low ASA values there is almost no catalytic activity. Sample A, which does not fit the correlation was not truly a carbon but more a semicoke, since its highest heat treatment temperature was only 700°C. Before measurement of ASA, this sample was heat treated at 950°C for 3 h under vacuum to clean the surface. By this treatment, sample A underwent a structural change resulting in a lower ASA, whereas the rate constant of NO reduction was measured without previous heat treatment. Sample B, prepared from sample A by heat treatment at 900°C for 1 h before measurement of the rate constant,

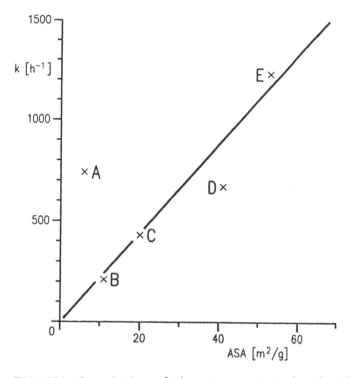

FIG. 19a Correlation of the rate constant for the NO reduction with the active surface area (ASA).

has a low rate constant as well as a low ASA and fits the correlation as can be expected.

Sample C was prepared from sample A by steam gasification to nearly 60 wt.% burn-off at 900°C, whereas sample D was an activated carbon with about 15 wt.% burn-off. A higher activity of sample D was expected because, the catalytic activity for NO reduction has a maximum at about 10-15 wt.-% burn-off and then decreases with increasing degree of burn-off as shown before in Section IVA. Sample D is the precurser of sample E, which was enriched with 5 wt.-% nitrogen and fits the correlation too.

Later, another correlation was made, taking into account that the rate constant for the NO reduction was based on volume of catalyst (vpm NO converted, related to residence time and catalyst bulk volume

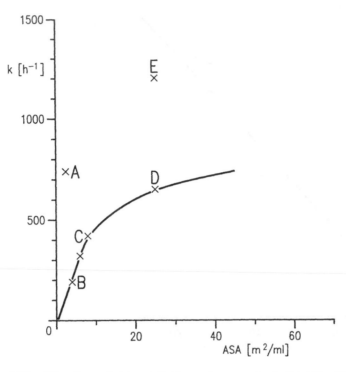

FIG. 19b Correlation of the rate constant for the NO reduction with
the ASA both based on bulk volume of catalyst.

$(k[h^{-1}]))$ and the ASA based on bulk volume of the catalyst too. The
measured values are shown in Fig. 19b and it can be seen that there
is no linear relationship at all, but the rate constant values are
running to a limiting value with increasing ASA. By comparison with
Fig. 15 (incorporation of nitrogen, Section IVA) there was found the
same behavior. The catalytic activity also rises with increasing
N-content, but not linearly, however reaching a limiting value for
a certain amount of incorporated nitrogen. Concerning the relationship between catalytic activity and ASA for samples without incorporated N (Fig. 19b) it can be seen, that the carbon catalysts D shows
the highest possible catalytic activity received by certain experimental conditions in the framework of the production process
described in Section IIb using a given precursor material.

The trend of the curves in Fig. 15 and Fig. 19b seems to have a general physical significance. The increase of the ASA or the amount of nitrogen beyond the limiting value of catalytic activity creates only active sites which have nearly no catalytic effect. Therefore the distribution of active sites of ASA against their catalytic activity is of great importance. This catalytic activity is related to the activation energy of the rate determining step of the catalytic reaction.

Suuberg et al. [37] have published a new model of a distributed activation energy concerning the ASA. The distribution of active centers against the activation energy can be described by a physically based mathematical model

The behavior of sample E in Fig. 19b corresponds to that of the catalytic activity of N incorporated samples related to the mean random reflectance (Fig. 18, Section IVB). This can be explained in terms of Boehm [10]. The incorporated nitrogen acts only as a promotor for catalytic oxidation reactions and does not lead to an increase of the amount of active centers (ASA). Later there are other promotors like halogens and/or transition metal oxides to increase the catalytic activity of carbon catalysts for the NO-reduction by a given value of ASA.

V. SULFUR DIOXIDE CONVERSION INTO SULFURIC ACID

A. Pore Structure and Quality of Active Sites

1. Mechanism of Reaction

For the adsorption of sulfur dioxide and its catalytic conversion into sulfuric acid the following consecutive reactions are assumed:

$$SO_2 \rightarrow SO_2(ads) \qquad H_2O \rightarrow H_2O(ads) \qquad \tfrac{1}{2}O_2 \rightarrow O(ads)$$
$$SO_2(ads) + O(ads) + H_2O(ads) \rightarrow H_2SO_4(ads) \tag{7}$$

In a first step, SO_2, O_2, and water are adsorbed, whereby the dependence of reaction rate on the square root of the oxygen concentration indicates that the oxygen adsorption takes place as a dissociative step [24]. In a consecutive reaction these adsorbed species react to sulfuric acid in a catalytic reaction. The product H_2SO_4 remains in an adsorbed state at the carbon surface even at temperatures around 200°C.

To clarify the mechanism of this consecutive reaction, break-through curves of three different catalysts were measured. For this experiment, (1) activated carbon, (2) active coke made from bituminous coal, and (3) peat coke [25] were used. For testing these catalysts, a gas mixture containing SO_2, O_2, H_2O, and N_2 was passed through a fixed bed filled with an equal volume of catalyst maintaining a con-stant gas residence time. The breakthrough of SO_2 as a function of time was recorded. As a result, typical breakthrough curves of the three different carbon adsorbents are shown in Fig. 20. The break-through curves of SO_2 sorption can be divided into three subsequent phases, in which phase change as a function of time is dependent on the properties of the catalysts.

In *phase I* (I in Fig. 20) no breakthrough is observed and the sorption rate is controlled by diffusion and adsorption of SO_2 at the internal surface of the catalysts. As the adsorption proceeds, the internal surface becomes occupied. Now vacancies on the internal

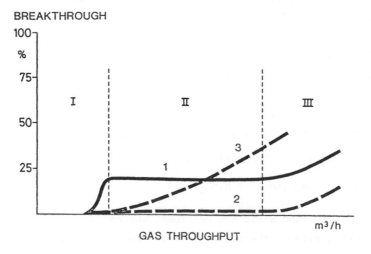

FIG. 20 Typical sulfur dioxide breakthrough curves for (1) activated carbon, (2) active coke (both made from bituminous coal), and (3) peat coke.

surface to allow continued adsorption can be created by catalytic
oxidation of adsorbed SO_2 to sulfuric acid and subsequent transport
of the generated sulfuric acid to readily accessible inner pores,
which are not adsorption sites for SO_2. Therefore, the sorption
rate in *phase II* (II in Fig. 20) is controlled by the rate of the
catalytic oxidation reaction, which itself is dependent on the quality
of active sites for oxidation and the pore structure of the catalyst.
It is possible that only the transport of sulfuric acid from the active
sites for SO_2 is controlling this reaction step. For the active coke
the rate of oxidation is as high as that of diffusion, for the investigated activated carbon the oxidation rate is distinctly lower, and for
the peat coke no steady state of the rate of oxidation can be reached
in phase two. In *phase III* (III in Fig. 20) the storage capacity of
the accessible pores for the sulfuric acid is occupied, the presence
of sulfuric acid begins to poison the adsorption sites available for
SO_2, and the adsorption activity declines. It can be seen that phase
three begins at the same time for active coke and activated carbon,
but begins much earlier for peat coke. Therefore, peat coke must
have less pores accessible for the storage of sulfuric acid.

   In general, it can be assumed that the carbon catalyst must have
different sites for the adsorption of SO_2 and for the storage of
sulfuric acid. The storage capacity for sulfuric acid may depend
mainly on pore structure of the adsorbent and may increase with
increasing burn-off of micropores. The active sites for SO_2 adsorption may correlate with the ASA; however, there exist no corresponding measurements. Therefore it is assumed that the pregraphitic
structure of active coke and that of related activated carbons have
active sites of the same quality but of different amount.

2.   *Influence of Particle Size and Pore Structure*

For optimization of catalyst performance it is important to know
more of the influence of the particle size and pore structure. The
pore structure can be varied systematically by a partial steam gasification of active coke toward well-defined stages of burn-off (see
Section IIB). A result of the interaction of particle size and pore

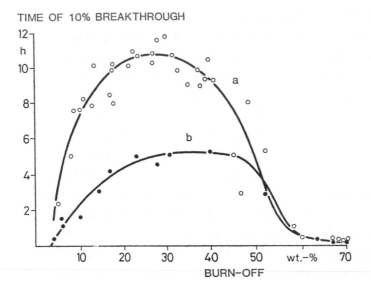

FIG. 21 Sulfur dioxide sorption of activated carbons made from
bituminous coal as a function of burn-off. Particle size: (a)
1-2 mm, (b) 4-6 mm.

structure as parameters influencing diffusion of SO_2 and storage of
H_2SO_4 is shown in Fig. 21 [26]. Here the time in hours after which
10% of the inlet sulfur dioxide will pass through the fixed bed
without being sorbed is related to the burn-off of carbon catalysts
under investigation. The particle size (1-2 mm and 4-5 mm) is
varied. It can be seen that there is a distinct maximum of catalytic activity as a function of burn-off for both grain fractions
at about 20-30 wt.% burn-off. At this burn-off the ratio of the
active internal surface that catalyzes the chemical reaction and the
pore volume for the storage of sulfuric acid formed by oxidation
seems to be optimal. Further, it follows from the figure that for
all catalysts with burn-off below 50 wt.% the fraction with a smaller
particle size is more active than with the larger one. Therefore, in
this area of burn-off, transport phenomena such as diffusion of SO_2
are rate-determining, at least for the products with the greater
particle size. Here the rate of diffusion is slower than that of the
chemical oxidation reaction.

3. Effect of Chemical Promoters

It follows that the application of additional active sites acceler-
ating the chemical reaction is only reasonable for small particle
sizes or in the area of burn-off beyond 50 wt.%. Under these conditions the rate of the chemical reaction can be increased by doping
the internal surface with typical oxidation catalysts, e.g., oxides
of cobalt, chromium, and vanadium. Some results of doping a product
of 60 wt.% burn-off and a grain size of 1-2 mm with metal oxides are
summarized as follows [26].

The addition of cobalt and vanadium oxides is only efficient if
the mixture of oxides and carbon is heated up to temperatures of
about 800°C after the impregnation from aqueous solution. This means
that the thermal reaction between the metal oxide and the carbon is
necessary to promote the catalytic efficiency.

In the case of cobalt oxide, a maximum effect as a function of
the amount of oxidic component can be observed at concentrations
between 1 and 2 wt.%. Therefore, it can be assumed that the active
oxide sites interact with the pregraphitic crystallites of the carbon
support. Most efficient is the addition of vanadium, potassium, and
chromium in the ratio 1:1:1. The catalyst shows its highest activity
after heating up to 600°C and exhibits constant activity during many
adsorption-regeneration cycles performed at a temperature of at maximum 600°C. In contrast, the activity without metal oxides declines
over several adsorption-regeneration cycles. It is supposed that
the decrease of activity in the absence of metal oxides is created
by the formation of surface oxide complexes at the efficient carbon
surface. These are decomposed fully in the presence of metal oxides
at 600°C, but are only partly decomposed at this temperature in the
absence of metal oxides. Therefore, an important effect of metal
oxides promoting catalytic activity is their ability to decompose
surface oxide complexes at low temperatures.

A comparison of the catalytic activity of activated carbon with
and without catalytic impregnation is shown in Fig. 22 [27]. Here
the burn-off is taken as parameter to characterize the carbon catalyst [18]. In both cases the activity increases initially with

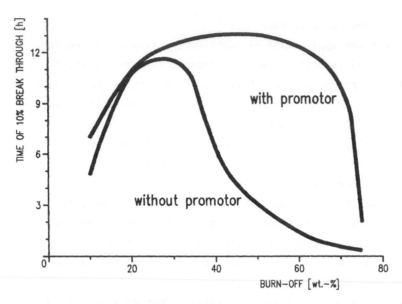

FIG. 22 Comparison of sulfur dioxide sorption of activated carbons with and without metal oxide impregnation as a function of burn-off.

increasing burn-off and then declines after reaching a maximum at about 25-30 wt.% and 50-60 wt.% burn-off, respectively.

It is interesting to see that the maximum of SO_2 uptake for the doped activated carbon is shifted to higher values of burn-off. However, the uptake value at the maximum is nearly the same for impregnated and nonimpregnated activated carbons (with a grain size of 1-2 mm). From a scientific point of view it can be assumed that the ASA decreases with increasing burn-off, and that the chemical promoters change the quality of active centers to higher catalytic activity even at higher burn-offs. The rate of diffusion controlling adsorption at higher burn-off was not checked further.

From the economic point of view this small difference in performance for products with high burn-off is not sufficient to use impregnated carbon as a catalyst commercially for three reasons: the impregnation is too costly compared with its effect, the mechanical strength of a catalyst with a burn-off of 60 wt.% is considerably less than that of a catalyst with a burn-off of 25%, and the metal

oxides also catalyze the oxidation of carbon, a reaction that is
not desirable.

4.   Comparison of NO/NO₃/C Reaction with the SO₂/C Reaction

4.   *Comparison of NO/NO₃/C Reaction with the
     SO₂/C Reaction*

As can be seen from Section III, the $NO/NH_3/C$ reaction is a catalytic
gas-solid reaction in which the reaction rate is controlled by the
chemical reaction and no mass transport restriction could be observed.
This was confirmed first by laboratory experiments using different
grain sizes at constant conditions and second by calculations [17].
After a short time a steady state is reached, and performing the
reaction in a fixed-bed reactor a constant concentration profile can
be observed as shown in Figs. 10-12. Optimal catalysts are activated
carbons with a very small burn-off. Relationships between catalytic
activity and properties of the catalyst such as reflectance and ASA
are known. Also, an interaction of certain chemical promoters with
the carbon surface is known [28]; however, their effect is very complex, and corresponding experiments are not yet finished to clarify
the mechanisms. This reaction can be performed in an industrial
fixed-bed reactor.

In contrast, the SO_2/C reaction leads to a final product of
the catalytic reaction—sulfuric acid—which remains adsorbed and
poisons the active sites of the carbon catalyst; therefore the reaction rate becomes smaller with time. From this it follows that no
steady state of reaction rate can be reached in a fixed-bed reactor,
because the reaction front is moving slowly through the reactor and
also the concentration profile is changed with time.

As shown in Fig. 20 the rate-determining steps of the overall
reaction are different as a function of time. In *phase I* the pore
diffusion of SO_2 may be rate-determining, especially using a carbon
catalyst with small burn-off and large grain size. In *phase II* the
oxidation reaction of SO_2 to sulfuric acid and/or the diffusion of
sulfuric acid from the active sites into the pore volume may be rate-
determining, and in *phase III* the rate is controlled by the poisoning
of active sites by the sulfuric acid adsorbed. Optimal catalysts are

activated carbons with a burn-off between 25 and 30 wt.%. In terms
of the active sites, there is little knowledge of their nature. However, the interaction of chemical promoters with the carbon surface
has been studied and the role of promoters in destroying surface
oxides at lower temperatures has been identified.

Because in a fixed-bed reactor no steady state of the concentration profile is reached, it is practical to use moving-bed reactors for industrial performing the SO_2/C reaction. From this moving-
bed reactor the loaded part of the catalyst is drawn out and must be
regenerated by a thermal process. The regenerated catalyst can be
given back to the top of the reactor. The thermal regeneration is
described in Section VB.

B.   Thermal Regeneration of the Sulfuric Acid-Loaded Catalysts

1.   *Mechanism*

At temperatures above 200°C active coke or activated carbon containing sulfuric acid undergoes a chemical reaction

$$H_2SO_4 + \tfrac{1}{2}C \rightarrow \tfrac{1}{2}CO_2 + H_2O + SO_2 \tag{8}$$

in which the adsorbed sulfuric acid reacts with carbon with formation
of CO_2, H_2O, and SO_2, which can be desorbed easily from the surface
of the carbon adsorbents.

This reaction has been investigated for a long time with intent
to determine kinetics, to transfer the laboratory results to commercial equipment, and to elucidate the reaction mechanism, which is
quite a bit more complex than shown in the simple reaction of Eq. (8).

To obtain suitable conditions to evaluate the reaction progress
according the well-known nonisothermal reaction kinetics [29], the
sulfuric acid-containing product is heated up at a constant rate of
heating of, for example, 1-10°C/min, and the volume of evolving individual reaction products as H_2O, SO_2, CO, and CO_2 is monitored versus
the change of time or temperature. For this treatment different
reactor types can be used, such as a differential flush reactor
purged with an inert gas in which no secondary reactions are observed,

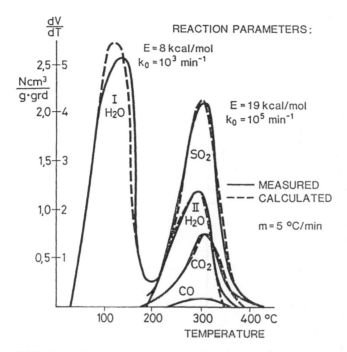

FIG. 23 Temperature-programmed desorption spectrum of active coke loaded with sulfuric acid, measured in a flush reactor at a heating rate of 5 K/min.

or a batch reactor with gas circuit to study consecutive reactions between products primarily evoluted. A typical result of a treatment in a differential flush reactor is shown in Fig. 23 [30]. Under the conditions of this experiment the regeneration reaction starts at around 200°C and is practically completed at 450°C, as indicated by the evolution of the reaction products.

The reaction parameters of frequency factor K_0 and activation energy E_A for the individual product reactions according to the Arrhenius law

$$r = K_0 \, e^{(-E_A/RT)} \qquad (9)$$

can be evaluated from the product formation peaks, taking into account the rules of nonisothermal reaction kinetics.

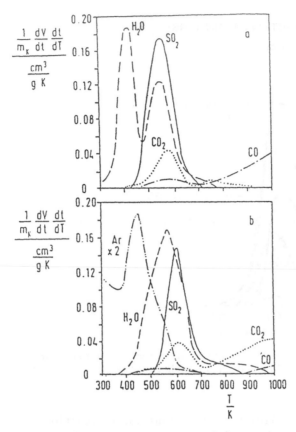

FIG. 24 Temperature-programmed desorption spectra for active coke
from a flue gas cleaning experiment, performed (a) in a differential
flush reactor and (b) in a batch reactor using a heating rate of
10 K/min.

It is interesting to consider the changes in reaction progress
as a function of temperature in a differential flush and in a batch
reactor with a gas circuit as used in most industrial reactor types
shown in Fig. 24 [31]. For the batch reactor the water vapor, the
sulfur dioxide peak, and all the other peaks of desorbed substances
are observed at higher temperatures compared to those of the peaks
for the flush reactor. Furthermore, the amounts of CO_2 and CO
evolved at temperatures above 700°C are different. These differences can be explained by the presence of higher concentrations of

$$H_2O_{ads} \rightleftharpoons H_2O$$

$$H_2SO_{4ads} \longrightarrow H_2O + SO_{3ads}$$

$$H_2SO_4 \cdot H_2O_{ads} \longrightarrow 2 H_2O + SO_{3ads}$$

$$SO_{3ads} \xrightarrow{C} SO_2 + \text{C-OXIDE (I)}$$

$$\text{C-OXIDE I} \longrightarrow CO_2$$

$$\text{C-OXIDE I} \rightleftharpoons \text{C-OXIDE II}$$

$$\text{C-OXIDE II} \longrightarrow CO$$

FIG. 25 Simplified reaction scheme of thermal regeneration of active coke loaded with sulfuric acid.

gaseous components in the batch reactor and corresponding consecutive reactions. Both results are very important for the design of industrial reactors.

The reaction scheme of Fig. 25 has been derived step by step, starting with single reactions and proceeding to more complex reactions. First a decomposition of sulfuric acid into water and SO_3 still in adsorbed state takes place. The latter reacts with the carbon, forming SO_2 and a surface oxide complex (I), which on the one hand immediately decomposes to CO_2 and on the other hand is converted into the complex (II), which forms only CO during decomposition at higher temperatures. The formation of surface oxide complexes of different properties (C oxide I and C oxide II) can also be established by their effect on catalytic cycles at different temperatures.

2. *Transfer to Higher Heating Rates*

In commercial regeneration units such as a moving bed, fluidized bed, or moving bed of a hot carrier in which the catalyst is fed, the heating rates are very different. Therefore the behavior of the reaction must be investigated at different heating rates. The temperature range at which a reaction with given parameters (K_0, E_A) proceeds is a function of the heating rate, as shown in Fig. 26 [27]. The calculations are based on the parameters established in laboratory experiments at low heating rates. It can be seen that the temperature range of the

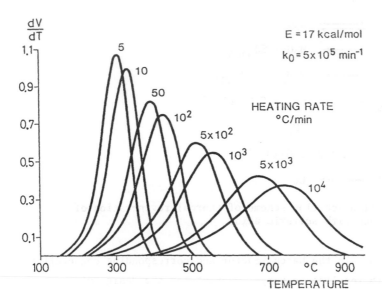

FIG. 26 Release of sulfur dioxide using different heating rates calculated from the reaction parameters E and k_0 determined at low heating rates.

decomposition reaction increases with increasing heating rate. The precondition is that the reaction mechanism remains unchanged in the different reactor types used for the performance or regeneration.

a. *Interaction of sulfur dioxide removal and regeneration in adsorption-regeneration cycles.* The flue gas must be cleaned continuously in the technical performance of SO_2 removal processes. For industrial application a moving-bed reactor with a flow of activated coke from top to bottom and cross-streamed by the flue gas is used in combination with a separate regeneration unit, in which the loaded active coke is regenerated and afterward fed back into the adsorption reactor. Sorption results for SO_2 taken from laboratory fixed-bed reactors can be extrapolated to industrial moving-bed reactors [32,33]. The change of properties of active coke during the course of adsorption-regeneration cycles in continuous operation has been systematically investigated [34]. The result was that pore volume and SO_2 sorption are increased during the first 80 cycles, while the mechanical strength

decreases slightly.   After 120 cycles a steady state without further
change of SO_2 sorption is reached.   Obviously an equilibrium is
established between the consumption of carbon by the chemical reaction
taking place during regeneration and mechanical attrition in the moving
bed on the one hand and the addition of fresh active coke on the other
hand.   The mean burn-off of the active coke in this steady state
corresponds to that of the maximum SO_2 sorption (see Fig. 21).

VI.   PERFORMANCE OF NH_3/NO REACTION IN THE PRESENCE OF SO_2

A.   Reaction Modeling on a Laboratory Scale

In Section VA the NH_3/NO/C reaction was compared with the SO_2/C
reaction.   It could be shown that the NH_3/NO/C reaction reaches a
steady state after a short time.   The SO_2/C reaction does not reach
a steady state, but a reaction front passes slowly through the fixed-
bed reactor.   Hoang Phu [35] has shown in his modeling of simultaneous
SO_2 and NO removal that in the part of a reactor without breakthrough
the SO_2/NH_3 reaction is faster than the NO/NH_3 reaction.   Figure 27
shows an experiment with a gas stream containing NO, NH_3, and SO_2, a

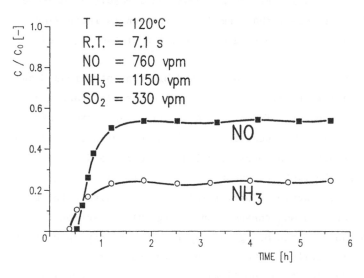

FIG. 27 Breakthrough curves of NO and NH_3 in the presence of SO_2
(Hoang Phu).

constant H_2O concentration (10% by volume), and a constant O_2 concentration (6% by volume) at an active coke as catalyst. Under these conditions the SO_2 removal is so fast that 100% conversion of SO_2 can be observed even after 6 h, while NO and NH_3 reach their steady-state conversions of 45% and 78%, respectively, after 30 min. For this case, in that SO_2 has not reached its breakthrough, one assumes the following reaction between SO_2 and NH_3:

$$SO_2 + 2NH_3 + H_2O + \tfrac{1}{2}O_2 \longrightarrow (NH_4)_2SO_4 \tag{10}$$

The reaction rate of this reaction was found to be

$$r_4 = \left(-\frac{dc_{SO_2}}{dt}\right) = k_6 c_{SO_2} c_{NH_3} \tag{11}$$

This equation does not describe the breakthrough concentration of SO_2 but is only valuable to describe the apparent steady state of SO_2 conversion in this part of the fixed-bed reactor, where no breakthrough of SO_2 is observed and in which not all active centers are occupied by sulfuric acid or $(NH_4)_2SO_4$, respectively (see Fig. 20, phases I and II). Due to the preferred NH_3 reaction with SO_2, the conversion of NO decreases. Figure 28 shows, as a result of measurement in a fixed-bed reactor, that the NH_3 conversion is increasing and the NO conversion is decreasing as a function of the SO_2 feed concentration.

Under the conditions of constant O_2 and H_2O concentrations Hoang Phu has assumed that there are four reactions in parallel taking place for the NO and SO_2 removal. The reaction equations and rate expressions are given in Fig. 29. Reaction 1 is identical with the reaction (1) used by Kleinschmidt. However reactions 2 and 3 are different. This may be due to the fact that another carbon catalyst was under investigation (e.g., reaction 2 only takes place at active coke, as can be seen by increased NH_3 consumption in Fig. 9). Reaction 3 was only assumed, because in the Hoang Phu experiments the oxygen content was constant and its influence was not investigated in relation to this reaction. Reaction 4 is identical to Eq. (11). The conditions

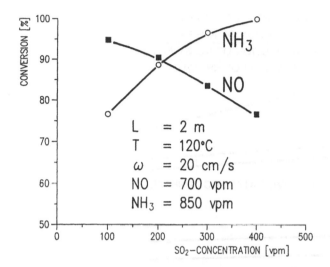

FIG. 28 Influence of the SO_2 feed concentration on NO conversion and NH_3 consumption (Hoang Phu).

PARALLEL REACTIONS:

(1) $\quad 2\,NO + 2\,C + O_2 \rightarrow N_2 + 2\,CO_2$
(2) $\quad 2\,NH_3 + 3/2\,O_2 \rightarrow N_2 + 3\,H_2O$
(3) $\quad 6\,NO + 4\,NH_3 \rightarrow 5\,N_2 + 6\,H_2O$
(4) $SO_2 + 2\,NH_3 + H_2O + 1/2\,O_2 \rightarrow (NH_4)_2 SO_4$

REACTION RATES MEASURED IN A FIXED BED REACTOR:

$$r_1 = \left(-\frac{1}{2}\frac{dc_{NO}}{dt}\right) = K_1\, c_{NO}{}^{0.6}$$

$$r_2 = \left(-\frac{1}{2}\frac{dc_{NH_3}}{dt}\right) = K_5\, c_{NH_3}$$

$$r_3 = \left(-\frac{1}{6}\frac{dc_{NO}}{dt}\right) = K_2\, \frac{c_{NO}\, c_{NH_3}}{(1 + K_3\, c_{NO} + K_4\, c_{NH_3})^2}$$

$$r_4 = \left(-\frac{dc_{SO_2}}{dt}\right) = K_6\, c_{SO_2} c_{NH_3}$$

FIG. 29 Reaction model of catalytic NO conversion by NH_3 at the active coke catalyst in the presence of SO_2 (Hoang Phu).

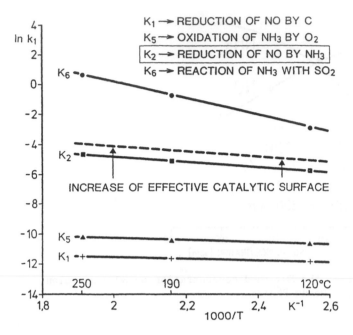

FIG. 30 Arrhenius plot of the parallel reactions relevant for the
NO/NH_3 reaction at the active coke catalyst in the presence of SO_2.

of validity of reaction are mentioned above. The total system fits
the measurements only in the restricted range of H_2O and O_2 concentrations, and at relatively short times related to the reaction 4
(no SO_2 breakthrough).

Plots of the reaction rate constants as a function of temperature are given in the Arrhenius diagram of Fig. 30 [36]. It can be
seen that the direct conversion of C with NO and oxygen into nitrogen
and carbon dioxide and the direct catalytic oxidation of ammonia in
the presence of carbon have minor significance compared with the
desired catalytic conversion of NO with NH_3, whose reaction rate
constant surpasses those of the other two reactions by five to seven
decades. This result corresponds with the result of Kleinschmidt.
The reaction rate constant and its temperature dependence for the
reaction of NH_3 with SO_2 to ammonium sulfate are even greater than
those for the conversion of NH_3 with NO. Therefore in an industrial
plant the overall reaction should be performed at the lowest possible

temperature to decrease the NH_3 consumption (as the activation energy
of the NH_3/SO_2 reaction is higher than that of the NH_3/NO reaction)
and possibly with low SO_2 concentrations.

Observations in fixed-bed reactors show that transport phenomena
may also be rate-determining related to the NH_3/SO_2 reaction, especially at high SO_2 concentrations. Therefore, the NH_3/H_2SO_4 reaction
takes place preferentially in a shell near the outer particle surface.
Under special reaction conditions ammonium sulfate can thus be created
at the outer grain surface, resulting in a conglomeration of grains.
The preferential performance of this reaction in the outer grain
regions has the advantage that no or only a few active sites responsible for the NO conversion with ammonia are deactivated. For the performance of both reactions in commercial units, the reaction conditions (temperature, SO_2 concentration, residence time of gas and
solid) have to be selected carefully.

As an example, in Fig. 31 the reaction model of Hoang Phu shown
in Fig. 30 is used for the calculation of the NO conversion in a pilot

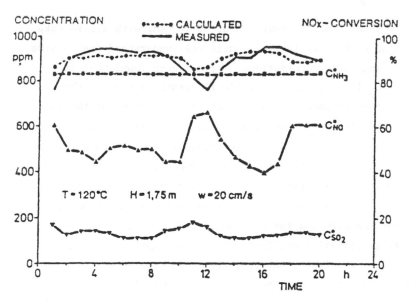

FIG. 31 Comparison of calculated and measured pilot-plant results
of NO_x removal in the presence of SO_2 (Hoang Phu).

plant with a feed stream containing NO, NH_3, and SO_2. It can be
seen that the calculated values fit the measured NO conversion very
well. The SO_2 conversion was very nearly 100% for the time of
measurement.

B. Regeneration of $(NH_4)_2SO_4$-Loaded Carbon Catalysts

The reaction of the decomposition of ammonium sulfate-loaded carbon
catalysts has been investigated in detail [31]. In principle the
decomposition follows the reaction scheme of Fig. 25. However,
because of the presence of ammonium compounds the reaction scheme is
more complex. Part of the nitrogen reacts to form NH_3, and this
ammonia is to some extent released at temperatures of about 600°C,
and is oxidized to some extent by surface oxides type I and type II
to elemental nitrogen at temperatures up to 700°C. Furthermore, part
of the nitrogen is fixed in the pregraphitic structure of the catalyst, increasing its catalytic activity for the $NO/NH_3/C$ reaction as
described in Section IVA.

VII. CONCLUSIONS

This review shows that the reaction of NO and NH_3 with active coke
and activated carbon catalysts is quite well understood over a wide
range of O_2 and H_2O concentrations. It has been found that the
presence of O_2 is necessary to enable the reaction through the
creation of the intermediate NO_2, which stays adsorbed at the carbon
surface. The reaction and reactor modeling make it possible to model
the system for a great variety of temperature and gas concentrations.
It has been proven that the Kleinschmidt model fits experimental
curves very well. From the scientific point of view it is interesting to see that the first step—the oxidation of NO to NO_2—is
the same for the C/NO and the $C/NO/NH_3$ reactions. In both cases
carbon acts as oxidation catalyst. The investigations have also
shown that the reactivity, which is chemically controlled, is related
to the active surface area, and there is an optimal carbon catalyst
consisting of an activated carbon with a very small burn-off. The

reactivity of very different carbons can also be related to their reflection. The system is applicable for NO_x removal for all flue gases in the absence of SO_2. In some cases—such as flue gases from gas-fired boilers, in which the oxygen concentration is nearly zero—an addition of O_2 is possible to enable the reaction.

The reactions responsible for the removal of SO_2 from flue gases are also very well understood. It is assumed that in different time-dependent phases different mechanisms are rate-determining, such as pore diffusion of SO_2, oxidation of SO_2, and/or transportion of sulfuric acid from active sites to inner pore volume and poisoning of active sites by H_2SO_4. From the scientific point of view it is important that an optimal activated carbon catalyst has 20-30 wt.% burn-off, and it follows that other active sites are more responsible compared with those for the $NO/NH_3/C$ reaction, or that the interaction of SO_2 adsorption at active sites with the transport of sulfuric acid from there in the inner pores is the precondition for the activity of the carbon catalyst. Possibly the relation between active sites and inner pores is the determining parameter. There are chemical promoters available for this reaction; however, their effect could only be observed at higher burn-off, and therefore these promoters have no industrial interest. In addition to the oxidation of SO_2 to sulfuric acid, the regeneration of sulfuric acid-loaded activated carbons can be transferred to an industrial scale. This process, however, is at this time not running in an industrial state, since this process of SO_2 removal cannot compete with washing processes.

The best application of the results of research is the simultaneous removal of SO_2 and NO_x. The influence of SO_2 inlet concentration and bed length (the latter at constant SO_2 inlet concentration) on the NO conversion is shown in Fig. 32 [16]. Figure 32 shows (right) that the NO conversion is sharply increased by decreasing SO_2 concentration at constant bed length, and (left) that even at SO_2 concentrations of 300 ppm high NO conversion can be reached with bed length up to 3 m. In order to achieve a high NO_x removal efficiency, a two-stage process design should be applied in an industrial plant

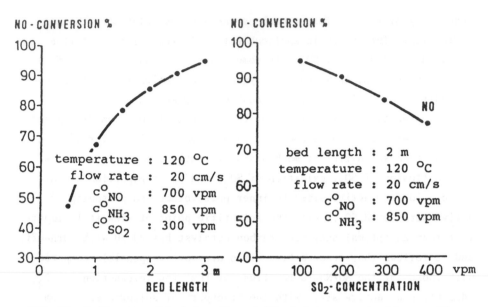

FIG. 32 Influence of bed length and SO_2 inlet concentration on NO conversion (Hoang Phu).

for the treatment of flue gases with high SO_2 concentrations. SO_2 must be removed in the first stage to about 100-200 ppm in the absence of ammonia to get a high NO conversion in the second stage even in the presence of these residual concentrations of SO_2. For this purpose moving-bed reactors are suitable in both stages. The reaction and reactor model of Hoang Phu, which fits the experimental conversion curves, can be used to calculate the NO conversion under narrow ranges of H_2O and O_2 concentrations and an apparent steady state of SO_2/C reaction. Large plants for the simultaneous removal of SO_2 and NO_x are running in Japan and are under construction in West Germany.

Finally, the authors wish to make two remarks. First, without doubt surface oxide complexes at the carbon catalysts play an important role in the reactions discussed in this chapter. For example, in the NO/C and the $NO/NH_3/C$ reaction the first step may be the adsorption of NO at a surface oxide and the formation of a $C-NO_2$ complex, which can react to further intermediates or final products with carbon, and with adsorbed NH_3 or H_2O. Furthermore, in the H_2SO_4/C reaction (regeneration of loaded carbon catalysts) two different surface oxides are formed, which release CO_2 or CO, respectively, during further heating. It seems that one or both surface oxides are destroyed by chemical promoters at lower temperature compared with the decomposition in absence of promoters. As known by experiments, the catalytic activity for adsorption of SO_2 and its oxidation to H_2SO_4 is inhibited by the presence of surface oxides, and in the way the discussed effect of promoters related to surface oxides can explain the increase of the chemical activity for SO_2 uptake in the presence of promoters. Here only some examples of the effect of surface oxides have been given; however, it was not possible to measure and to define the structures of surface oxides and their change during the reaction described. Therefore this cannot give a full impression of role of surface oxides, which will be a very important task for future research.

The second question is related to the reaction and reactor modeling described in this chapter by the Kleinschmidt and Hoang Phu models. It could be shown that both can be used to describe the $NO/NH_3/C$ reaction under distinct but different conditions, since it is possible to fit the measured points with calculated values by adjusting the reaction constants. However, it does not follow that these models describe the mechanisms of the reaction: they are only useful to describe the course of the reaction in larger reaction units.

REFERENCES

1. H. Jüntgen. Activated carbon as catalyst support—A review of new research results. *Fuel 65*:1436-1446 (1986).

2. N. Barranco. Dissertation, RWTH Aachen (1976).

3. H. O. Müller von Blumencron and H. P. Gelbke. Phosgen. In *Encyclopädie der technischen Chemie*, 4. Aufl., Bd. 18. Verlag Chemie, Weinheim, 1979, pp. 275-282.

4. G. Jonas and L. Mischke. Schwefel-Halogen-Verbindungen, anorganische. In *Ullmanns Enzyklopädie der technischen Chemie,* 4. Aufl., Bd. 18. Verlag Chemie, Weinheim, 1982, pp. 75-85.

5. K. D. Henning, J. Klein, and H. Jüntgen. *VDI-Forschungsheft 615*:35 (1983).

6. J. Klein and K. D. Henning. *Fuel 63*:1064-1067 (1984).

7. H. Jüntgen and E. Richter. Dokumentation Rauchgasreinigung, VDI-Verlag, Sept. 1985.

8. B. Tereczki, R. Kurth, and H. P. Boehm. *Reprints Carbon '80, 3rd Int. Carbon Conf.*, Baden-Baden, June 30-July 4, 1980, pp. 218-222.

9. T. Stöhr and H. P. Boehm. *Reprints Carbon '86, 4th Int. Carbon Conf.*, Baden-Baden, June 30-July 4, 1986, pp. 354-356.

10. A. Vass, T. Stöhr, and H. P. Boehm. *Reprints Carbon '86, 4th Int. Carbon Conf.*, Baden-Baden, June 30-July 4, 1986, pp. 411-413.

11. H. Yehaskel. *Activated Carbon, Manufacture and Regeneration.* Noyes Data Corporation, Park Ridge, NJ, 1978.

12. H. Kienle and E. Bäder. *Aktivkohle und ihre industrielle Anwendung.* Ferdinand Enke-Verlag, Stuttgart, 1980.

13. H. Jüntgen. *Carbon 6*:297 (1986).

14. H. Jüntgen. *Veröffentlichungen des Bereichs und Lehrstuhls fur Wasserchemie* Heft 9. Karlsruhe, 1975, pp. 23-35.

15. H. Jüntgen. *Gezielte Herstellung von Adsorptionskoksen für die Wasser- und Luftaufbereitung.* Habilitationsschrift, Heidelberg, 1966.

16. H. Jüntgen. Activated carbon as catalyst, Erdöl und Kohle, Erdgas. *Petrochemie/Hydrocarbon Technol.* 39:546-551 (1986).

17. R. Kleinschmidt. Dissertation, GHS Essen, 1988.

18. H. J. Mühlen. Frühjahrstagung "Arbeitskreis Kohlenstoff," Karlsruhe, 31 March 1987.

19. H. Kühl, E. Richter, K. Knoblauch, and H. Jüntgen. *Reprints Carbon '86, 4th Int. Carbon Conf.*, Baden-Baden, June 30-July 4, 1986, pp. 351-353.

20. A. Oberlin and J. N. Rouzand. *ACS Symposium Series* No. 303, *Petroleum Derived Carbons*, 1986, p. 85.

21. H. Jüntgen, E. Richter, and H. Kühl. Catalytic activity of carbon catalysts for the reaction of NO_x and NH_3. *Fuel 67*: 775-780 (1988).

22. H. Kühl, H. Baumann, H. Jüntgen, P. Ehrburger, J. Dentzer, and J. Lahaye. The importance of active surface area on the NO-reduction with ammonia on carbon catalysts. *Fuel 68*:129-130 (1988).

23. N. R. Laine, F. J. Vastola, and P. L. Walker Jr. *Phys. Chem.* 67:2030 (1963).

24. H. Dratwa, H. Jüntgen, and W. Peters. *Chem.-Ing. Technol. 39:* 949-955 (1967).

25. H. Dratwa and H. Jüntgen. *Staub-Reinh. der Luft 27*(7):301-307 (1967).

26. M. Kruel. *Brennst.-Wärme-Kraft 23*(3):91-97 (1971).

27. P. Steiner, H. Jüntgen and K. Knoblauch. *Advances in Chemistry Series 139*:180-191 (1975).

28. Patent, Hitachi, Ltd., Tokyo, 100904-74.

29. H. J. Schröter, H. Jüntgen, and W. Peters. *Carbon 11*(2):93-102 (1973).

30. K. H. van Keek and H. Jüntgen. *Ber. der Bunsenges. phys. Chem. 72*(9/10):1223-1231 (1986).

31. J. Jung, E. Richter, K. Knoblauch, and H. Jüntgen. *Chem. Eng. Fundam. 2*(1):39-52 (1983).

32. H. Jüntgen, K. Knoblauch, and M. Kruel. *Chem.-Ing.-Technol. 41*(2):77-81 (1970).

33. E. Richter, K. Knoblauch, and H. Jüntgen. *vt-Verfahrenstechnik 14*(5):338-342 (1980).

34. H. Jüntgen, K. Knoblauch, and D. Zündorf. *Chem.-Ing.-Technol. 45*(19):1148-1151 (1973).

35. T. Hoang-Phu. *Dissertation, GHS Essen, 1984.*

36. H. Jüntgen, E. Richter, K. Knoblauch, and T. Hoang-Phu. *Chem. Eng. Science 43*(3):419-428 (1988).

37. E. M. Suuberg, M. Wojowwicz, and J. M. Calo. Proceedings of an Int. Conf. on Carbon, Carbon '88, University of Newcastle upon Tyne, England, September 18-23, 1988, pp. 325-327.

3

Theory of Gas Adsorption on Structurally Heterogeneous Solids and its Application for Characterizing Activated Carbons

MIECZYSLAW JARONIEC

Institute of Chemistry, M. Curie-Sklodowska University, Lublin, Poland

JERZY CHOMA

Institute of Chemistry, Military Technical Academy, Warsaw, Poland

I. INTRODUCTION

Solid adsorbents of important industrial applications usually possess
a complex porous structure, which consists of pores of different sizes
and shapes. The significance of these pores in the adsorption process
depends on their sizes. In the case of larger pores (macropores and
mesopores) their surface is covered according to the layer-by-layer
adsorption mechanism [1]. These pores play also an important role in
the transport of adsorbate molecules inside of the adsorbent particles.
Fine pores (micropores) are a source of a substantial increase in the
adsorption capacity because their whole accessible volume may be
regarded as the adsorption space; adsorption in the micropores occurs
according to the volume-filling mechanism [2]. According to the IUPAC
classification of pore size [3], micropores are defined as pores with
widths not exceeding 2 nm, mesopores are pores with widths between 2
and 50 nm, and macropores are pores with widths exceeding 50 nm. This
classification is to some extent arbitrary, since the adsorption mechanism is dependent not only on the pores size but also on the size and
structure of the adsorbate molecule.

General use of porous solids in science and technology [4]
requires their many-faceted characterization. Various modern tech-

niques provide a direct information about physicochemical properties
of these solids. Although the role of these techniques for characterizing porous adsorbents is still increasing [5], the classical adsorption and desorption measurements are very popular and are still carried
out because they provide information about behavior of a solid with
respect to an adsorbate [6]. The adsorption-desorption isotherm is
most frequently used to calculate the specific surface area [7-10]
and the pore-size distribution [1,11], which are fundamental quantities recommended by the IUPAC for characterizing porous adsorbents.
The specific surface area evaluation requires only a small part of
the adsorption isotherm [7]. The pore-size distribution is evaluated
from either the multilayer part of the adsorption isotherm or porosi-
metric measurements based on the penetration of mesopores and macropores by liquid mercury; this distribution does not include micropores
[1].

Part of the adsorption isotherm measured at low concentrations has
not been explored sufficiently for studying adsorption processes and
for characterizing the microporous structure of the solid adsorbents [6].
This part of the adsorption isotherm is a source of valuable information about adsorbate-adsorbent interactions [12] and energetic and
structural heterogeneities of porous solids [6,13-16]. Significant
progress in the theoretical description of gas and vapor adsorption
on microporous solids [1,2,6,13-16] provides foundations for utilizing
the low-concentration adsorption measurements to evaluate the adsorption potential distribution and the micropore-size distribution, which
characterize energetic and structural heterogeneities of the micropores.

This review deals with the theory of gas and vapor adsorption on
heterogeneous microporous solids and its application for interpreting
the experimental isotherms and for characterizing physicochemical
properties of these solids. Although the presented theoretical considerations have a general character, their experimental verification
is only shown for adsorption on activated carbons, which are adsorbents of great industrial importance [17-19].

II.   SELECTED ASPECTS OF ADSORPTION ON POROUS SOLIDS

A.   Adsorption in Micropores and Mesopores

Adsorption on a porous solid occurs in micropores, mesopores, and
macropores. Usually the amount adsorbed on the macropore surface is
negligible in comparison to that occuring in micropores and mesopores
[20]. At the low relative pressures the micropores are occupied by
the adsorbate molecules according to the volume-filling mechanism
[2]. The superposition of the adsorption forces generated by the
opposite walls of the micropores causes a significant increase in the
adsorption potential inside them [21,22]; consequently, the amount
a_{mi} adsorbed in the micropores gives a main contribution to the total
adsorbed amount a_t [20]. At moderate relative pressures the micropore
filling process is accompanied by layer-by-layer adsorption on the
mesopore surface. At high relative pressures capillary condensation
occurs in the mesopores. According to Dubinin [16,20], the total
adsorbed amount a_t is given by

$$a_t = a_{mi} + a_{me} \tag{1}$$

where a_{me} is the amount adsorbed on the mesopore surface. Equation
(1) is utilized for extracting the amount a_{mi} adsorbed in the micropores from the total adsorbed amount a_t.

B.   Adsorption Methods for the Assessment of Microporosity

The most popular methods used for the assessment of microporosity are
based on a comparison of the shape of a given isotherm with that of
a standard isotherm on a nonporous reference solid [1,23-25]. Two of
them, the t-method and α_s-method, are frequently used for analyzing
adsorption isotherms on microporous solids [26-36]. In the t-method
of Lippens and de Boer [26], the adsorbed amount is plotted against
t, the corresponding multilayer thickness calculated from the standard
isotherm. This method was modified by Sing [27] to provide a method
for the assessment of microporosity. Limitations of this method, connected with evaluation of the statistical film thickness from the
standard isotherm, are discussed elsewhere [23].

In 1970 Sing [31] proposed the α_s-method for analyzing adsorption isotherms on microporous solids. In this method, the amount adsorbed a_t is plotted against the reduced standard adsorption α_s, which is defined for a nonporous reference adsorbent as the ratio of the amount adsorbed a_r at the relative pressure p/p_s to the amount a_r^s adsorbed at $p/p_s = s$; usually s is assumed to be equal to 0.4. A theoretical formulation of the α_s-method may be made on the basis of Eq. (1) [37]. For larger values of the relative pressure p/p_s the micropores are filled; in this case, the first term in Eq. (1) is constant and equal to the micropore adsorption capacity a_{mi}^0. Then Eq. (1) assumes the following form:

$$a_t = a_{mi}^0 + a_{me} \qquad (2)$$

After filling the micropores, adsorption occurs on the mesopore surface according to the layer-by-layer formation of the multilayer. If the reference nonporous solid surface possesses the same physico-chemical properties as the mesopore surface of the adsorbent studied, then for the same value of the relative pressure p/p_s the relative adsorptions θ_r and θ_{me} are identical, that is,

$$\theta_r = \theta_{me} \quad \text{where} \quad \theta_r = a_r/a_r^0 \quad \text{and} \quad \theta_{me} = a_{me}/a_{me}^0 \qquad (3)$$

Here a_r^0 and a_{me}^0 denote the monolayer capacities of the reference nonporous solid and the mesopore surface, respectively. Combination of Eqs. (2) and (3) gives

$$a_t = a_{mi}^0 + a_{me}^0 \theta_r \qquad (4)$$

Note that for $\theta_r > 1$ the symbol θ_r denotes the number of the statistical adsorbed layers, which multiplied by the thickness of one layer gives the multilayer thickness t; this thickness is utilized in the above mentioned t-method.

Combining the definition of $\alpha_s = a_r/a_r^s$ with Eq. (4), we obtain an equation for describing the linear part of the α_s-plot:

$$a_t = a_{mi}^0 + \eta\alpha_s \quad \text{where} \quad \eta = a_{me}^0 a_r^s/a_r^0 \qquad (5)$$

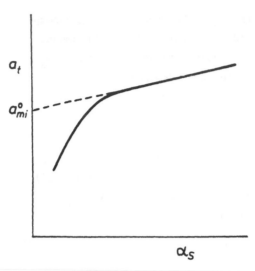

FIG. 1 Schematic representation of the α_s plot.

A schematic illustration of the α_s-plot is shown in Fig. 1; as with
the t-method, the back extrapolation of this plot provides the micropore adsorption capacity a_{mi}^0 as the intercept on the adsorption axis.
This micropore capacity may be converted to a liquid volume. It is
noteworthy that the choice of a suitable reference material for applying the α_s-method as well as t-method is an important problem and was
the subject of recent debate [34,38].

Dubinin and Kadlec [39,40] proposed the so-called t/θ_{mi}-method
for initial analysis of the adsorption isotherms on microporous
solids. According to this method the amount adsorbed a_t is plotted
against the quantity t/θ_{mi}, where t is the statistical film thickness
and θ_{mi} represents the relative adsorption in the micropores, that
is, $\theta_{mi} = a_{mi}/a_{mi}^0$. Although the t/θ_{mi}-plot should be linear in a
wider region than the linearity region of the α_s-method, the evaluation of the quantity t/θ_{mi} requires additional assumptions, which
limit the applicability of the t/θ_{mi}-method [41]. This method is used
frequently by Dubinin and co-workers [2,42-46].

Another method used for the assessment of microporosity is preadsorption [47-52], in which large molecules (e.g., nonane) are used

to fill the micropores; these molecules are not removed by pumping the adsorbent at ambient temperature. This method can provide an effective way of isolating the micropores and leaving the mesopore surface available for the adsorption of other adsorbates [3]. Measurements of the adsorption isotherms for an adsorbate (e.g., nitrogen) on carbonaceous samples without and with preadsorbed nonane permit their quantitative comparison by using the isotherm subtraction method [48,49,51]; this method seems to be promising for analyzing microporous structures of activated carbons.

C.   Evaluation of the Monolayer Capacity and Surface Area of the Mesopores

It follows from a brief presentation of the t-method and α_s-method (cf. Section IIB) that the intercept on the adsorption axis evaluated by the back extrapolation of the t-plot or α_s-plot provides the micropore adsorption capacity a_{mi}^0 (cf. Fig. 1). The slope of the linear segment of the t-plot or α_s-plot permits evaluation of the monolayer adsorption capacity a_{me}^0 for the mesopore surface. In the α_s-method, this linear segment is described by Eq. (5); since a_r^s and a_r^0 are known for a given standard adsorption isotherm, the value of a_{me}^0 may be calculated from η. The specific surface area S_{me} of the mesopores is obtained by multiplying a_{me}^0 by Avogadro's number N_A and the molecular area ω occupied by one molecule adsorbed on the mesopore surface, that is,

$$S_{me} = a_{me}^0 N_A \omega \tag{6}$$

The t-method and α_s-method provide an effective and simple way for evaluating the mesopore surface area S_{me} as well as the micropore adsorption capacity a_{mi}^0. For the purpose of illustration, Figs. 2 and 3 present the α_s-plots for the benzene adsorption isotherms on selected activated carbons at 293 K; the experimental details concerning the adsorption measurements are given elsewhere [14,53-56]. Table 1 contains the basic information about activated carbons and references to the experimental isotherms of benzene adsorption. The values of a_{mi}^0, a_{me}^0, and S_{me} for the benzene adsorption isotherms

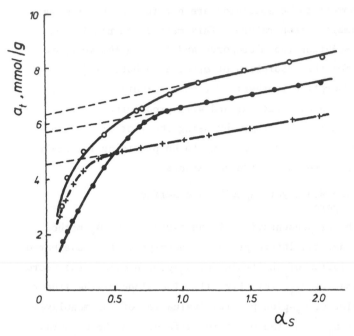

FIG. 2 The α_s plots for the benzene adsorption isotherms on activated carbons CWZ-3 (white circles), AC (black circles), and RKD-4 (crosses) at 293 K.

from Table 1, obtained by plotting them according to the α_s-method (cf. Figs. 2 and 3), are summarized in Table 2. This table contains results of the α_s-method for the selected adsorption systems, which were extensively studied in earlier papers [37,57,58]. Note that the α_s-plots were made by using the benzene adsorption isotherm at 293 K measured on a carbon black, which was obtained from the active furnace soot (Podkarpacka Rafinery, Poland) by heating at 1173 K in an argon atmosphere [37,59]; this isotherm was applied to calculate the reduced standard adsorption α_s. The BET parameters evaluated from this standard isotherm in the range of p/p_s from 0.06 to 0.35 are $a_r^0 = 0.32$ mmol/g, $C_{BET} = 38.9$, and $S_{BET} = 79$ m^2/g. At $p/p_s = 0.4$ the amount adsorbed $a_r^s = 0.49$ mmol/g [37].

It is noteworthy that the parameters of the α_s-method depend on the standard adsorption isotherm; therefore the choice of a suitable

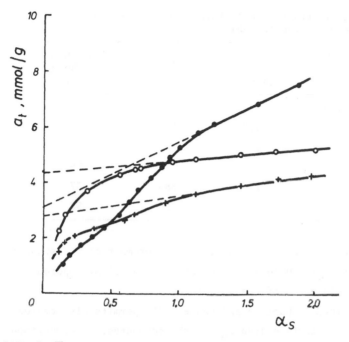

FIG. 3 The α_S plots for the benzene adsorption isotherms on activated carbons F-400 (white circles), BH (black circles), and HY (crosses) at 293 K.

TABLE 1 Basic Information about the Benzene Adsorption Isotherms on Activated Carbons at 293 K

Code of activated carbon	Source of activated carbon	Reference for the adsorption isotherm
CWZ-3	Comp. of Carbon Electrodes, Raciborz, Poland; obtained from plant products by two-step activation process	53
AC	Merck, West Germany; commercial product	54
RKD-4	Norit, The Netherlands; commercial product	55
F-400	Chemviron, USA; Filtrasorb 400 obtained from coal by high-temperature steam activation	56
BH	Bender-Hobein AG, Zurich, Switzerland; commercial product	14
HY	Lurgi-Bayer, West Germany; Hydraffin 71	56

TABLE 2   Parameter Values of the α_s-Method for the Benzene
Adsorption Isotherms Listed in Table 1

Code of activated carbon	a_{mi}^0 (mmol/g)	a_{me}^0 (mmol/g)	S_{me} (m^2/g)	Reference for the α_s parameters
CWZ-3	6.30	0.68	166	37
AC	5.67	0.59	150	57
RKD-4	4.53	0.58	140	57
F-400	4.33	0.30	75	58
BH	3.11	1.55	383	37
HY	2.82	0.48	120	58

reference nonporous carbon and an accurate measurement of the standard
isotherm are important problems in characterization of microporous
activated carbons in terms of the α_s-method.

The t/θ_{mi}-method of Dubinin and Kadlec [40] permits also evaluation of the specific surface area S_{me} of the mesopores. A comparison
of this method with other methods is presented elsewhere [41,60]. It
follows from these comparative studies [41,60] that the specific mesopore surface area S_{me} may be also evaluated by means of the following
methods, e.g., preadsorption method [47,52], method based on estimation of the area of the film adsorbed on the mesopore surface [16,61],
methods utilizing the mesopore-size distribution [62-64], and method
utilizing measurements of the immersion enthalpy [60,65]. These
methods provide sometimes different values of S_{me} [41,60], and therefore special care is recommended in selecting a suitable method for
evaluating the specific surface area of the mesopores. The IUPAC
recommendations [3] and extensive experimental studies [31-37,57,58]
show that the α_s-method provides a simple and attractive way for
evaluating the specific surface area of the mesopores.

D.   Extraction of the Amount Adsorbed in the Micropores
     from the Total Adsorbed Amount

Equation (1) permits extraction of the amount a_{mi} adsorbed in the
micropores from the total adsorbed amount a_t. It follows from Section IIB that if a_{me}^0 is known, then the adsorbed amount a_{me} may be

estimated on the basis of the standard adsorption isotherm θ_r, that is, $a_{me} = a_{me}{}^0\theta_r$. Then the amount a_{mi} is

$$a_{mi} = a_t - a_{me}{}^0\theta_r \tag{7}$$

An alternative to Eq. (7) was proposed by Dubinin [2,16,20]:

$$a_{mi} = a_t - S_{me}\gamma_r \tag{8}$$

where γ_r is the amount adsorbed per unit surface area of the reference nonporous solid, that is, $\gamma_r = a_r/S_r$, where S_r is the specific surface area of the reference solid. Equations (7) and (8) indicate that extraction of a_{mi} from the total adsorbed amount a_t requires knowledge of the monolayer capacity $a_{me}{}^0$ for the mesopore surface or the specific surface area S_{me} of the mesopores; these quantities may be evaluated according to the α_s-method or other methods discussed briefly in Section IIC. Equation (8) was frequently used by Dubinin and co-workers [16,40,42-45,66,67] for calculating the adsorption in the micropores.

III. TOTAL ENERGETIC HETEROGENEITY OF POROUS SOLIDS

A. Concept of Energetic Heterogeneity for Porous Solids

The porous structure of activated carbons is complex. These carbons possess slit-like micropores of different sizes and forms, as well as the differentiated mesopores [1,2]. The micropores and mesopores of different sizes are source of structural heterogeneity, which may be described by the pore-size distribution function. While this pore-size distribution may be relatively easily evaluated in the range of the mesopores and macropores [1], its calculation in the range of the micropores is difficult. On the other hand, the micropore-size distribution provides the most valuable information about structural heterogeneity of activated carbons because the micropores change their sorption properties substantially and thus their adsorption capacities; therefore, by the structural heterogeneity of activated carbons, we usually understand heterogeneity of the microporous structure, which plays a dominant role in the adsorption process at low and moderate relative pressures.

Activated carbons with a large specific surface area of the
mesopores, besides the structural heterogeneity of the micropores,
possess a significant surface heterogeneity. The mesopore surface
possesses irregularities, imperfections, strongly bound impurities,
and various functional groups, which are source of the surface heterogeneity [6,18,19]. The total adsorbent heterogeneity of activated
carbons comprises their structural and surface heterogeneities, while
the structural heterogeneity of these solids may be described by the
micropore-size distribution function, their surface heterogeneity is
more difficult for a quantitative description.

Adsorption measurements are the source of information about
adsorbate-adsorbent interactions, which may be characterized by the
adsorption energy or adsorption potential [6,12,13]. The adsorption
energy distribution $X_t^*(U)$ is a generally accepted function for a
quantitative characterization of the global energetic heterogeneity
of a solid [6,13,68-73]. This distribution function describes the
so-called "relative heterogeneity," i.e., energetic heterogeneity of
a solid with respect to a given adsorbate [73]. The function $X_t^*(U)$
provides only information about distribution of the number of adsorption sites with respect to their adsorption energy U, but it does not
inform us about the source of the energetic heterogeneity; therefore,
additional independent investigations are required to determine the
physicochemical nature of this energetic heterogeneity [6,73]. In
the case of microporous solids, a comparison of the total energy distribution function with that for a reference nonporous solid permits
a separation of the contributions to the total distribution function
$X_t^*(U)$ arising from the structural and surface heterogeneities [14,74].

B. Integral Equation for the Total Adsorption Isotherm
    and Evaluation of the Energy Distribution Function

The fundamental integral equation for the total adsorption isotherm
may be written as follows [6,13,69,73]:

$$\theta_t = \int_\Omega \theta(p,U) X_t^*(U) \, dU \tag{9}$$

where $\theta_t = a_t/a_t^0$ is the relative adsorption, $a_t^0 = a_{mi}^0 + a_{me}^0$
defines the total adsorption capacity, $\theta(p,U)$ is the isotherm

equation describing the local adsorption on the sites of the adsorption energy U, Ω is the integration region, and the energy distribution function $X_t^*(U)$ is normalized to unity.

Many attempts have been made to solve the integral of Eq. (9) with respect to the $X_t^*(U)$ function [6,13,69-73, and references therein]. The most advanced numerical methods for evaluating the energy distribution function $X_t^*(U)$ from the total adsorption isotherm $\theta_t(p)$ were discussed and compared by Jaroniec and Brauer [6]. It follows from their report [6] that the regularization methods of solving the integral Eq. (9) with respect to $X_t^*(U)$ give fair promise to be useful for determining the energetic heterogeneity of solids from the experimental adsorption isotherms. Although these methods were already used for calculating the energy distribution function $X_t^*(U)$ for nonporous and wide-porous solids [71,75-78], they were applied only in a few cases to evaluate $X_t^*(U)$ for microporous activated carbons [79]; further studies in this direction are desirable.

A simple and effective way for evaluating the energy distribution function from the total adsorption isotherm creates the condensation approximation method [70]. The accuracy of this method is quite satisfactory at low temperatures, and it increases when the adsorption temperature decreases [7]. This method provides a simple equation for calculating the total adsorption potential distribution $X_t(A)$ [70]:

$$X_t(A) = -d\theta_t(A)/dA \qquad (10)$$

where the adsorption potential A is defined as follows [2]:

$$A = RT \ln(p_s/p) \qquad (11)$$

In the condensation approximation method, the adsorption potential distribution $X_t(A)$ has the same shape as the distribution function $X_t^*(U)$; these functions are both shifted relative to each other on the energy axis only [80].

C. Total Adsorption Potential Distribution Associated
   with the Exponential Adsorption Isotherm

Based on Eq. (10), Jaroniec [81] proposed a simple method for calculating the total adsorption potential distribution $X_t(A)$ from the

experimental adsorption isotherm. According to this method the
adsorption isotherm is described by the following logarithmic poly-
nomial:

$$\ln a_t = - \sum_{j=0}^{N} B_j [RT \ln(p_s/p)]^j \quad \text{for} \quad p \leq p_s \tag{12}$$

where $B_j = 1,2,\ldots,N$ are the approximation coefficients. Equation
(12) corresponds to the exponential equation for the characteristic
adsorption curve $\theta_t(A)$:

$$\theta_t = a_t/\exp(-B_0) = \exp \left[- \sum_{j=1}^{N} B_j A^j \right] \tag{13}$$

An attempt at the derivation of Eq. (13) in terms of statistical
thermodynamics was made elsewhere [81]. It is noteworthy that for
special sets of the B_j parameters Eq. (13) may be reduced to the
classical Freundlich and Dubinin-Radushkevich (DR) equations [81].

The adsorption potential distribution $X_t(A)$ associated with
Eq. (13) is obtained by differentiating this equation according to
Eq. (10):

$$X_t(A) = \left[\sum_{j=1}^{N} j B_j A^{j-1} \right] \exp \left[- \sum_{j=1}^{N} B_j A^j \right] \tag{14}$$

Equation (14) has been applied to calculating the distribution function
$X_t(A)$ for many adsorption systems [6]. This equation was also used to
calculate $X_t(A)$ for carbonaceous adsorbents [14,15,82].

Equation (14) contains a large number of adjustable parameters
B_j for $j = 1,2,\ldots,N$, so its great flexibility allows more complex-
shaped distributions $X_t(A)$ to be determined; for instance, a dual peak
distribution can be obtained [81]. However, the majority of the adsorption potential distributions $X_t(A)$ obtained from the adsorption data by
means of Eq. (14) show one asymmetrical peak with a great widening in
direction of the higher values of A [14,15,82]. To illustrate the
shape of $X_t(A)$ we present in Fig. 4 the adsorption potential distributions $X_t(A)$ for benzene adsorbed on activated carbons (AC) (solid line),
RKD-4 (dashed line), and HY (dotted line), which were selected from

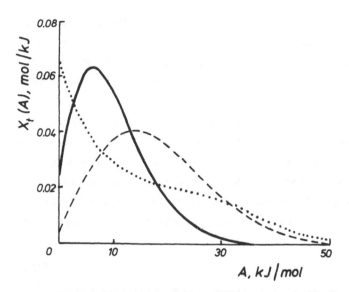

FIG. 4 Adsorption potential distribution $X_t(A)$ calculated according to Eq. (14) for benzene adsorbed on activated carbons AC (solid line), RKD-4 (dashed line), and HY (dotted line) at 293 K. The parameter N = 3 (carbon AC), 2 (carbon RKD-4), and 3 (carbon HY).

Table 1. Evaluating the adsorption potential distribution $X_t(A)$ by means of a method that utilizes the curve-fitting procedure [e.g., Eq. (12)] we should remember that the significant sensitivity of this distribution to changes in experimental adsorption data requires caution in its physical interpretation, particularly when a limited number of experimental points is available.

IV. ADSORPTION ON HOMOGENEOUS MICROPOROUS SOLIDS

A. Dubinin-Radushkevich Equation

The Dubinin-Radushkevich (DR) equation, proposed in 1947 [83], undoubtedly occupies a central position in the theory of physical adsorption of gases and vapors on microporous solids [2]. According to this equation the amount a_{mi} adsorbed in the micropores is a simple exponential function of the square of the adsorption potential A:

$$a_{mi} = a_{mi}^{0} \exp[-B(A/\beta)^2] \qquad (15)$$

Here B is the temperature-independent structural parameter that is associated with the micropore sizes, and β is the similarity coefficient, which reflects the adsorbate properties [2]. The DR equation, Eq. (15), is commonly used for studying gas and vapor adsorption on microporous activated carbons, and this problem is well known in the adsorption literature [1,2,84]. In these studies benzene is used as a reference adsorbate and then $\beta = 1$ [2].

An important stage in the theory of gas adsorption on microporous solids was determination of the relationship between the structural parameter B and the half-width x of the slit-like micropores. Experimental studies of Timofeev [85], Stoeckli [86], and Dubinin [87] showed that for the particular case of carbonaceous adsorbents with slit-like micropores with limited lateral dimensions, B is proportional to the square of x:

$$B = \zeta x^2 \tag{16}$$

where ζ is the proportionality constant; for benzene adsorbed on activated carbons this constant was estimated to be equal to 0.00694 mol/kJ nm)^2 [15,16]. Combining Eqs. (15) and (16), Dubinin [16] transformed Eq. (15) to the following form:

$$a_{mi} = a_{mi}^0 \exp(-mx^2 A^2) \tag{17}$$

where

$$m = \zeta/\beta^2 \tag{18}$$

Based on extensive experimental studies of adsorption on microporous activated carbons, Dubinin [87] and Stoeckli [88] postulated that the DR equation [Eq. (15)] describes adsorption in uniform micropores; this postulate permitted elaboration of the theory of gas and vapor adsorption on heterogeneous microporous solids, which will be presented later.

The condensation approximation method [70] gives the following adsorption potential distribution function $X_{mi}(A)$ associated with the DR equation [Eq. (15)]:

$$X_{mi}(A) = -d\theta_{mi}(A)/dA = 2BA\beta^{-2} \exp[-B(A/\beta)^2] \tag{19}$$

where $\theta_{mi} = a_{mi}/a_{mi}{}^0$. This distribution reflects the situation that even inside uniform micropores the adsorption force field may be nonuniform. Although the function $X_{mi}(A)$ given by Eq. (19) contains the error produced by the condensation approximation method [70], it characterizes energetic heterogeneity of the adsorption space inside micropores [89].

The shape of the distribution function $X_{mi}(A)$ given by Eq. (19) is well known to be an asymmetrical peak with a widening in the direction of high values of A. This function reaches the maximum at the point $A_m = \beta/(2B)^{1/2}$. The average adsorption potential $\overline{A}$ associated with Eq. (19) is given by [90]

$$\overline{A} = (\beta/2)(\pi/B)^{1/2} \tag{20}$$

The dispersion σ_A associated with Eq. (19) is defined as follows:

$$\sigma_A = (1 - \pi/4)^{1/2}\beta/\sqrt{B} = (4/\pi - 1)^{1/2}\overline{A} \tag{21}$$

A characteristic feature of the adsorption potential distribution $X_{mi}(A)$ given by Eq. (19) is the constant ratio of σ_A to $\overline{A}$, that is, $\sigma_A/\overline{A} = 0.523$ [90].

B. Langmuir-Freundlich Equation

Twenty years before the first formulation of Eq. (15) [83], Chakravarti and Dhar [91] proposed an empirical equation that gives a good representation of many experimental adsorption isotherms, especially isotherms measured on activated carbons [92-98]. This equation is reducible to the Langmuir and Freundlich (LF) equations [13] and written in terms of the adsorption potential A has the following form [99]:

$$a_{mi} = a_{mi}{}^0\{1 + \exp[\nu(A - \overline{A})]\}^{-1} \tag{22}$$

where ν is the temperature-independent parameter and $\overline{A}$ is the average adsorption potential of the distribution $X_{mi}(A)$ associated with Eq. (22); this distribution is given by

$$X_{mi}(A) = \nu \exp[\nu(A - \overline{A})]/\{1 + \exp[\nu(A - \overline{A})]\}^2 \tag{23}$$

Several papers provided the theoretical foundations for Eq. (22). Sips [100] derived the energy distribution function associated with

Eq. (22) by inverting the integral of Eq. (9). Bering and Serpinsky [101,102] derived the LF equation in terms of the vacancy solution model of adsorption; they also showed that this equation may be used for describing adsorption on microporous solids. Dubinin et al. [103] combined these treatments and concluded that the source of the surface-phase nonideality predicted by the LF equation is the energetic hetero-geneity of the adsorbent.

C. Comparison of DR and LF Equations

An important step in the theoretical studies of adsorption on micro-porous solids was the recognition of the analogy between DR Eq. (15) and LF Eq. (22). Dubinin's studies [104,105] led him to the conclu-sion that these equations are equivalent. In further papers [106, 107] he and his co-workers found approximate relationships between the parameters of Eqs. (15) and (22). This idea was developed successfully by Jaroniec and Marczewski [108,109], who found general relationships between parameters of these equations and proved that they show similar behavior over a wide region of pressures. These studies were also continued and verified experimentally [57,99].

It follows from the comparison of Eqs. (15) and (22) that for $A = \overline{A}$, Eq. (15) gives $\theta_{mi} = 0.456$, whereas Eq. (22) gives $\theta_{mi} = 0.5$. In the limiting case, for $A = 0$, Eq. (15) gives $\theta_{mi} = 1$, whereas Eq. (22) assumes the finite value of $\theta_{mi} = (1 + \exp(-\nu\overline{A}]^{-1}$, which is smaller than unity. Since for $\nu RT = 1$ Eq. (22) reduces to the Lang-muir equation, it shows the Langmuir behavior at the high-pressure region, that is, $\theta_{mi} \to 1$ when the equilibrium pressure tends to infinity. The comparative studies of Jaroniec and Marczewski [108] led to the following relationships between parameters of Eqs. (15) and (22):

$$B = (\beta\nu)^2/8 \tag{24}$$

$$a_{mi}^{0,DR} = 0.824 \, a_{mi}^{0,LF} \tag{25}$$

Equation (24) is especially important because it permits expression of the parameter ν by means of the half-width x [99]:

$$\nu = \kappa x \quad \text{where} \quad \kappa = (2/\beta)(2\zeta)^{\frac{1}{2}} \tag{26}$$

Equation (26) was obtained by combining Eqs. (16) and (24).

For the purpose of illustration, the benzene adsorption isotherms listed in Table 1 were described by Eqs. (15) and (22); Table 3 contains the values of the parameters associated with these equations and the values of the percentage relative error ε. Experimental studies [57] showed that LF Eq. (22) gives slightly better description of the adsorption data than Eq. (15). Figure 5 presents the dependence of a_{mi} versus A plotted according to Eqs. (15) (solid line) and (22) (dashed line) for the benzene adsorption on activated carbons CWZ-3 (white circles) and BH (black circles). It follows from this figure that these equations both describe satisfactory benzene adsorption on the CWZ-3 carbon. A poorer description is observed for benzene adsorption on the BH carbon; it will be shown later that this carbon possesses a strong structural heterogeneity and therefore Eqs. (15) and (22) give a poor representation of the adsorption data.

Figure 6 shows the functions $X_{mi}(A)$ associated with the curves presented in Fig. 5 for activated carbon CWZ-3. Other details concerning the comparison of Eqs. (15) and (22) are published elsewhere [57].

TABLE 3 Values of the Parameters of Eqs. (15) and (22) for the Benzene Adsorption Isotherms Listed in Table 1[a]

Code of carbon	DR Eq. (15)			LF Eq. (22)			
	a_{mi}^{0} (mmol/g)	$\overline{A}$ (kJ/mol)	ε (%)	a_{mi}^{0} (mmol/g)	$\overline{A}$ (kJ/mol)	ν (mol/kJ)	ε (%)
CWZ-3	5.55	19.9	5.52	6.99	15.3	0.13	3.61
AC	4.84	14.0	11.74	5.99	11.6	0.21	12.31
RKD-4	4.89	18.9	2.36	5.20	17.4	0.16	2.11
F-400	3.99	20.0	8.83	6.03	11.7	0.11	4.17
BH	1.65	14.1	11.58	2.30	10.1	0.19	11.10
HY	2.30	22.5	1.41	2.92	16.8	0.11	0.94

[a]Note: Third column of this table contains $\overline{A}$ associated with the DR Eq. (15), which through Eq. (20) permits calculation of B.

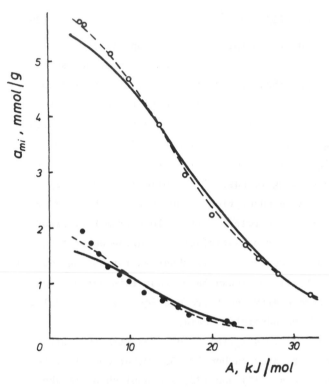

FIG. 5   Amount a_{mi} adsorbed in the micropores plotted as a function
of the adsorption potential A according to Eq. (15) (solid lines) and
Eq. (22) (dashed lines) for the benzene adsorption on activated carbons
CWZ-3 (white circles) and BH (black circles) at 293 K.

To illustrate the analogy between Eqs. (15) and (22) we plotted
the dependence of $a_{mi}^{0,DR}$ versus $a_{mi}^{0,LF}$ for data from Table 3 (Fig.
7); the solid line represents the theoretical line given by Eq. (25).
However, Fig. 8 shows the dependence of B^{DR} versus B^{cal}, which was
calculated according to Eq. (24).   Illustrative calculations (cf.
Figs. 5-8) confirmed that DR Eq. (15) and Eq. (22) show analogous
behavior over a wide pressure range, and this guarantees correlation
of the adsorption parameters according to Eqs. (24) and (25).   It is
noteworthy that the observed equivalence of the LF Eq. (22) (which
has a rigorous thermodynamical foundation) and the DR Eq. (15) provides new theoretical arguments for the DR equation, which as a semi-
empirical character.

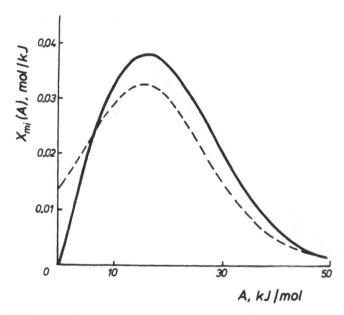

FIG. 6 Adsorption potential distribution $X_{mi}(A)$ calculated according to Eq. (19) (solid line) and Eq. (23) (dashed line) for benzene adsorbed on activated carbon CWZ-3 at 293 K.

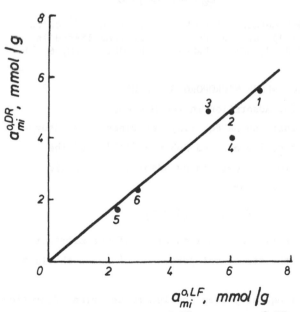

FIG. 7 Dependence of $a_{mi}^{0,DR}$ versus $a_{mi}^{0,LF}$ for the adsorption systems listed in Table 3; the solid line is plotted according to Eq. (25). Code: 1, CWZ-3; 2, AC; 3, RKD-4; 4, F-400; 5, BH; 6, HY.

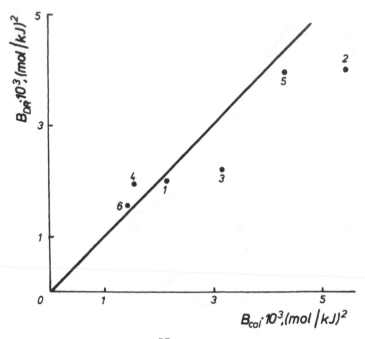

FIG. 8 Dependence of B^{DR} obtained from DR Eq. (15) on B^{cal} calculated according to Eq. (24) for the adsorption systems listed in
Table 3.   Code: 1, CWZ-3; 2, AC; 3, RKD-4; 4, F-400; 5, BH; 6, HY.

V.   ADSORPTION ON HETEROGENEOUS MICROPOROUS SOLIDS

A.   Integral Equations for Adsorption in the Micropores

Based on the postulate that the DR Eq. (15) describes only adsorption in uniform micropores, Izotova and Dubinin [110] used the
following two-term equation for describing adsorption on solids
with bimodal microporous structure:

$$\theta_{mi} = f_1 \exp[-B(A/\beta)^2] + f_2 \exp[-B_2(A/\beta)^2] \qquad (27)$$

Here f_i is the volume fraction of the i-th class of the micropores
($f_1 + f_2 = 1$) and B_i is the structural parameter of the i-th class
of the micropores.

Equation (27) has frequently been applied to describe adsorption
on nonuniform microporous solids [2,16,55,87,111-114].   The model
studies [55] showed that Eq. (27) is especially suitable for describing adsorption on microporous solids that possess two types of micro-

pores of considerably different sizes. For microporous solids with a great number of micropores of different sizes, the summation in Eq. (27) should be replaced by integration, and then θ_{mi} is given by [88]

$$\theta_{mi} = \int_0^\infty \exp[-B(A/\beta)^2]F(B) \ dB \tag{28}$$

where $F(B)$ is the distribution function of the structural parameter B normalized to unity. The integral Eq. (28) was first proposed by Stoeckli [88]. Expressing in this integral the structural parameter B by means of the half-width x [cf. Eq. (16)] we have [16]

$$\theta_{mi} = \int_0^\infty \exp(-mx^2A^2)J(x) \ dx \tag{29}$$

where $J(x)$ is the micropore-size distribution and m is defined by Eq. (18).

Comparison of the integral Eqs. (28) and (29) gives the following relationship between the distribution functions $F(B)$ and $J(x)$ [112]:

$$J(x) = 2\zeta x F[B(x)] \quad\quad \text{where} \quad\quad B = \zeta x^2 \tag{30}$$

Another form of the integral Eq. (28) was also considered by Dubinin and Kadlec [46].

B. Integral Equation for the Adsorption Potential
   Distribution Associated with the Micropore-Size
   Distribution

Differentiation of the integral Eqs. (28) and (29) with respect to A gives equations for the adsorption potential distribution $X_{mi}(A)$ [89, 115]:

$$X_{mi}(A) = 2A\beta^{-2} \int_0^\infty B \ \exp[-B(A/\beta)^2]F(B) \ dB \tag{31}$$

and

$$X_{mi}(A) = 2A\beta^{-2}m \int_0^\infty x^2 \ \exp(-mx^2A^2)J(x) \ dx \tag{32}$$

Equations (31) and (32) define the relationship between the distribution function $X_{mi}(A)$, $F(B)$, and $J(x)$. The average adsorption potential $\overline{A}$ associated with Eqs. (31) and (32) is given by [115]

$$\overline{A} = \frac{\beta\sqrt{\pi}}{2} \int_0^\infty \frac{F(B)}{\sqrt{B}} \, dB = \frac{1}{2}\left(\frac{\pi}{m}\right)^{\frac{1}{2}} \int_0^\infty \frac{J(x)}{x} \, dx \tag{33}$$

The dispersion σ_A for the distribution function $X_{mi}(A)$ may be expressed as follows [115]:

$$\sigma_A = \left[\beta^2 \int_0^\infty \frac{F(B)}{B} \, dB - \overline{A}^2\right]^{\frac{1}{2}} + \left[\frac{1}{m} \int_0^\infty \frac{J(x)}{x} \, dx - \overline{A}^2\right]^{\frac{1}{2}} \tag{34}$$

where $\overline{A}$ is defined by Eq. (33).

Equations (33) and (34) have general character and permit calculation of $\overline{A}$ and σ_A for arbitrary micropore distribution functions $F(B)$ and $J(x)$.

C. Analytical Solutions of the Integral Equations for Adsorption in the Micropores

In this section we will present briefly the analytical equations for θ_{mi} derived on the basis of the integral Eqs. (28) and (29). Stoeckli [88] solved the integral Eq. (28) for Gaussian distribution $F(B)$:

$$F(B) = [\sigma_B(2\pi)^{\frac{1}{2}}]^{-1} \exp[-(B - \overline{B})^2/2\sigma_B^2] \tag{35}$$

where $\overline{B}$ is the average value of B and σ_B is the dispersion of Gaussian distribution

Integration of Eq. (28) for Gaussian distribution function $F(B)$ gives [88]

$$\theta_{mi} = \exp[-\overline{B}(A/\beta)^2] \exp[(A/\beta)^4 \sigma_B^2/2]\{0.5[1 - \text{erf}(W)]\} \tag{36}$$

where

$$W = [(A/\beta)^2 - \overline{B}/\sigma_B^2]\sigma B/\sqrt{2} \tag{37}$$

There is an inconsistency in the derivation of Eq. (36). The distribution function $F(B)$ given by Eq. (35) satisfies the normalization condition in the interval $(-\infty, +\infty)$, whereas Eq. (28) was integrated over the interval $(0,\infty)$; in consequence, the isotherm Eq. (36) was derived for the "truncated" Gaussian distribution F(B) [that is, Eq. (35) for B $\geq$ 0], which does not satisfy the normalization condition

$$\int_{0}^{\infty} F(B)\ dB = 1 \tag{38}$$

To satisfy this condition, the Gaussian distribution should be multiplied by the factor $2/[1 + \text{erf}(\overline{B}/\sigma_B\sqrt{2})]$ [116]; consequently, the isotherm Eq. (36) should be also multiplied by this factor. This factor is negligible for $\overline{B}/\sigma_B > 2.3$ because the erf function is close to unity. Another disadvantage of the "truncated" Gaussian distribution [that is, Eq. (35) for B $\geq$ 0] is its nonzero value for B = 0, which is physically unrealistic; B = 0 implies that the micropore dimension x = 0 and then the distribution function should be equal to zero [54]. Stoeckli et al. [117-122] and Dubinin et al. [87,112-114] showed that Eq. (36) gives a good representation of many adsorption isotherms on activated carbons.

Rozwadowski and Wojsz [116,123] solved the integral Eq. (28) for distribution functions F(B) other than the Gaussian one, e.g., decreasing and increasing exponential distributions and Rayleigh distribution. Some of these equations have rather complex mathematical form; however, the decreasing exponential and Rayleigh distributions may be considered as special cases of the gamma-type distribution, which generates a very simple isotherm equation for θ_{mi} [89]; this equation will be discussed later.

Wojsz and Rozwadowski [124-126] also studied the adsorption potential distributions $X_{mi}(A)$ associated with the above-mentioned distribution functions F(B); these distributions were evaluated by differentiating equations for the characteristic adsorption curve $\theta_{mi}(A)$ with respect to A, that is, $X_{mi}(A) = -d\theta_{mi}(A)/dA$. It is noteworthy that they may be also obtained on the basis of the integral Eq. (31).

Dubinin [16] recommended studying the gas adsorption on heterogeneous microporous solids on the basis of the integral Eq. (29); this integral contains the micropore-size distribution, which gives direct

information about structural heterogeneity of the micropores.   In
his paper [16] Dubinin proposed a Gaussian distribution to represent
the micropore size distribution $J(x)$:

$$J(x) = [\sigma_x (2\pi)^{1/2}]^{-1} \exp[-(x - \bar{x})^2 / 2\sigma_x^2] \tag{39}$$

where $\bar{x}$ is the average value of x and σ_x is the dispersion of Gaussian
distribution.

Integration of Eq. (29) for the Gaussian micropore-size distribution $J(x)$ gives [16]

$$\theta_{mi} = [2Y(A)]^{-1} \exp\left\{-\frac{m\bar{x}^2 A^2}{[Y(A)]^2}\right\}\left[1 + \mathrm{erf}\left(\frac{\bar{x}}{\sigma_x \sqrt{2}\, Y(A)}\right)\right] \tag{40}$$

where

$$Y(A) = (1 + 2m\sigma_x^2 A^2)^{1/2} \tag{41}$$

The isotherm Eq. (40) shows a similar inconsistency to Eq. (36).   The
adsorption potential distribution $X_{mi}(A)$ associated with Eq. (40) was
discussed elsewhere [74,127].   Equation (40) was extensively studied
by Dubinin et al. [16,42,44-46,66,128-131]; it was used for describing the benzene adsorption on various types of microporous activated
carbons.   These studies showed its better applicability for describing
the adsorption data on microporous solids in comparison to Eq. (36).

The integral Eq. (29) has also been solved for micropore-size
distributions $J(x)$ other than the Gaussian one [132], but the obtained
isotherm equations are complex and less useful for interpreting the
adsorption data on microporous solids.

D.   Isotherm Equation Associated with the Gamma
     Micropore-Size Distribution

Jaroniec and Choma [15] applied a very simple equation for representing the adsorption isotherm θ_{mi}:

$$\theta_{mi} = [q\beta^2 / (q\beta^2 + A^2)]^{n+1} \tag{42}$$

Here $q > 0$ and $n > -1$ are the parameters of the gamma-type distribution function $F(B)$:

$$F(B) = \frac{q^{n+1}}{\Gamma(n+1)} B^n \exp(-qB) \tag{43}$$

Equation (42) was obtained by integrating Eq. (28) with the function F(B) given by Eq. (43); this function has a maximum at $B_m = n/q$. The average value $\bar{B}$ and the dispersion σ_B associated with Eq. (43) are given by [115,133]

$$\bar{B} = (n + 1)/q \tag{44}$$

$$\sigma_B = (n + 1)^{\frac{1}{2}}/q \tag{45}$$

The micropore-size distribution $J(x)$ generated by Eq. (43) is also a gamma-type function [15]:

$$J(x) = [2(q\zeta)^{n+1}/\Gamma(n + 1)]x^{2n+1} \exp(-q\zeta x^2) \tag{46}$$

This function has a maximum at the point

$$x_m = [(n + \tfrac{1}{2})/(q\zeta)]^{\frac{1}{2}} \tag{47}$$

Let us introduce:

$$\phi_x = \frac{\Gamma(n + 3/2)}{(n + 1)^{\frac{1}{2}}\Gamma(n + 1)} \tag{48}$$

The gamma-type micropore-size distribution $J(x)$ given by Eq. (46) is characterized by the values of $\bar{x}$ and σ_x [115]:

$$\bar{x} = \phi_x[(n + 1)/(q\zeta)]^{\frac{1}{2}} \tag{49}$$

$$\sigma_x = [(1 - \phi_x^2)(n + 1)/(q\zeta)]^{\frac{1}{2}} \tag{50}$$

Experimental verification of Eq. (42) [15,37,53,57,58,133,134] showed that this equation frequently gives a better representation to the adsorption data on microporous solids than DR Eq. (15). To illustrate the utility of Eq. (42) for describing adsorption in the micropores, the benzene isotherms listed in Table 1 were analyzed in order to evaluate the parameters a_{mi}^0, q, and n. A comparison of the values of a_{mi}^0, evaluated by the α_s-method (cf. Table 2), DR Eq. (15) (cf. Table 3), and Eq. (42) (cf. Table 4) for the benzene adsorption isotherms, showed that all methods provide similar values for the micropore adsorption capacity. Figure 9 presents the experimental dependence of a_{mi} versus A for selected adsorption systems from Table 4; this dependence is well described by Eq. (42). Comparing this

TABLE 4 Values of a_{mi}^0, q, and n Evaluated According to Eq. (42) for Benzene Adsorption Isotherms Listed in Table 1

Code of carbon	a_{mi}^0 (mmol/g)	q (kJ/mol)^2	n	Reference for the parameters
CWZ-3	6.06	1277	2.50	37
AC	6.18	175	0.59	This work
RKD-4	4.92	2073	3.94	This work
F-400	4.45	728	1.29	58
BH	2.67	97	0.21	37
HY	2.42	1543	2.13	58

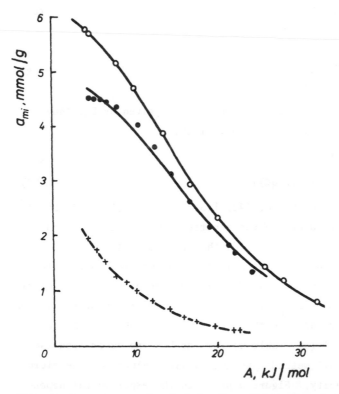

FIG. 9 Amount a_{mi} adsorbed in the micropores plotted as the function of A for benzene adsorbed on activated carbons CWZ-3 (white circles), RKD-4 (black circles), and BH (crosses). The solid lines were calculated according to Eq. (42).

description with that presented in Fig. 5 we see that especially for strongly heterogeneous microporous solids (e.g., carbon BH) Eq. (42) gives a better representation of adsorption data than DR Eq. (15).

The adsorption potential distribution $X_{mi}(A)$ associated with Eq. (42) is given by [15]

$$X_{mi}(A) = 2(n + 1)q^{n+1}(A/\beta^2)/[q + (A/\beta)^2]^{n+2} \tag{51}$$

The maximum of $X_{mi}(A)$ appears at the point

$$A_m = \beta[q/(2n + 3)]^{\frac{1}{2}} \tag{52}$$

Equations (33) and (34) permit derivation of equations for A and σ_A; integration of these equations for the gamma distribution function F(B) given by Eq. (43) leads to the following expressions [133]:

$$\overline{A} = \frac{\beta\phi_A}{2}\left(\frac{\pi q}{n}\right) \tag{53}$$

$$\sigma_A = \beta[(1 - \pi\phi_A^2/4)(q/n)]^{\frac{1}{2}} \tag{54}$$

where

$$\phi_A = \Gamma(n + \frac{1}{2})/[\sqrt{n}\,\Gamma(n)] \tag{55}$$

The quantities $\overline{A}$ and σ_A characterize the adsorption potential distribution $X_{mi}(A)$. Application of Eq. (42) for characterizing the microporous structure of activated carbons will be shown in the next section.

As we end this section we would like to mention that Eq. (42) was derived by integrating Eq. (28) over the interval $(0,\infty)$; this means that this equation is suitable for strongly heterogeneous microporous solids. For microporous solids possessing a weak structural heterogeneity the integral Eq. (28) should be defined in the interval (B_0,∞), where B_0 is the minimal value of B; then the following equation is obtained [89]:

$$\theta_{mi} = \exp[-B_0(A/\beta)^2][q\beta^2/(q\beta^2 + A^2)]^{n+1} \tag{56}$$

Studies of Eq. (56) showed that for many adsorption isotherms on microporous activated carbons, B_0 is very small and the exponential term of this equation approaches unity [54]; in consequence, Eq. (56) reduces to the simple Eq. (42).

VI. CHARACTERIZATION OF ACTIVATED CARBONS

A. Structural Heterogeneity of Micropores

The parameters q and n calculated according to Eq. (42) (cf. Table 4)
permit evaluation of the micropore distributions F(B) [that is, Eq.
(43)] and J(x) [that is, Eq. (46)], which characterize structural
heterogeneity of micropores. To illustrate the shape of the distribution functions F(B) and J(x), in Figs. 10 and 11 we present these
functions for the selected activated carbons. The distribution function F(B) shows asymmetry (cf. Fig. 11), which becomes smaller for
the micropore-size distrubition J(x) (cf. Fig. 11). Thus, for the
suitable values of the parameters q and n, Eq. (46) produces an
almost symmetrical function J(x), which shows similar behavior to
the Gaussian distribution given by Eq. (39); an additional advantage
of Eq. (46) in comparison to Eq. (39) is its behavior at x = 0. For

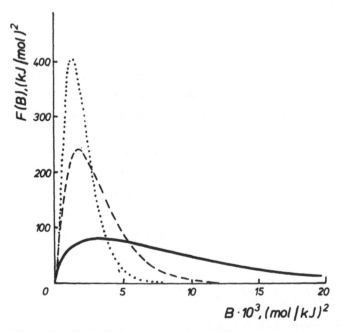

FIG. 10 Distribution function F(B) calculated according to Eq. (43)
for activated carbons AC (solid line), F-400 (dashed line), and HY
(dotted line).

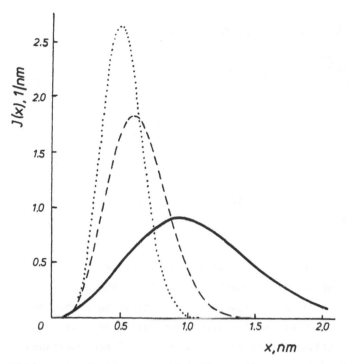

FIG. 11 Micropore-size distribution function J(x) calculated according to Eq. (46) for activated carbons AC (solid line), F-400 (dashed
line), and HY (dotted line).

x = 0 the distribution function J(x) is equal to zero and satisfied
the physical requirement.

Dubinin [135] used the quantities σ_x of the Gaussian distribution
[Eq. (39)] to characterize the structural heterogeneity of the micropores; these quantities are more useful in characterization of activated
carbons than a graphical presentation of the micropore-size distribution. In the previous section we discussed equations for $\bar{B}$, σ_B and
$\bar{x}$, σ_x associated with the gamma-type distribution functions F(B) and
J(x). Table 5 contains the values of $\bar{B}$, σ_B and $\bar{x}$, σ_x for adsorption
systems listed in Table 4; the parameters q and n required for their
calculation are summarized in Table 4. The quantities $\bar{x}$ and σ_x are
especially useful for comparing microporous structures of various
activated carbons. For example, Table 5 shows that the activated
carbon RKD-4 possesses the smallest structural heterogeneity (the

TABLE 5 Values of $\bar{B}$, σ_B and $\bar{x}$, σ_x for Adsorption Systems Listed in Table 4

Code of carbon	$\bar{B} \times 10^3$ [Eq. (44)] $(mol/kJ)^2$	$\sigma_B \times 10^3$ [Eq. (45)] $(mol/kJ)^2$	$\bar{x}$ [Eq. (49)] (nm)	σ_x [Eq. (50)] (nm)
CWZ-3	2.77	1.48	0.61	0.17
AC	9.09	7.21	1.06	0.43
RKD-4	2.38	1.07	0.57	0.13
F-400	3.15	2.08	0.64	0.22
BH	12.47	11.34	1.21	0.57
HY	2.03	1.15	0.52	0.15

smallest value of σ_x), whereas the microporous structure of carbon BH is strongly heterogeneous (the highest value of σ_x). The quantity $\bar{x}$ informs us about sizes of the micropores, e.g., activated carbon HY possesses more micropores of smaller sizes than carbon BH. Other details concerning activated carbons listed in Table 5 are published elsewhere [37,57,58,133]; the quoted papers also contain other examples of activated carbons, which were characterized by means of the quantities $\bar{x}$ and σ_x. The presented studies of activated carbons by means of Eq. (42) give a favorable forecast for its use in characterizing heterogeneity of their microporous structures.

B. Physical Interpretation of the Total Adsorption Potential Distribution for Microporous Solids

Equation (1) expressing the total adsorbed amount a_t may be rewritten as follows [74]:

$$\theta_t = f_{mi}\theta_{mi} + f_{me}\theta_{me} \tag{57}$$

where $\theta_{mi} = a_{mi}/a_{mi}^0$ and $\theta_{me} = a_{me}/a_{me}^0$ denote the relative adsorptions in the micropores and mesopores, respectively; $\theta_t = a_t/a_t^0$ is the total relative adsorption; $a_t^0 = a_{mi}^0 + a_{me}^0$ is the total adsorption capacity; and $f_{mi} = a_{mi}^0/a_t^0$ and $f_{me} = a_{me}^0/a_t^0$ are the fractions of the adsorption capacities for the micropores and mesopores, respectively. Note that the adsorption capacity a_{me}^0 is defined as the

monolayer capacity for the mesopore surface. The condensation
approximation method gives the following expression for the total
adsorption potential distribution $X_t(A)$ associated with Eq. (57)
[74]:

$$X_t(A) = f_{mi}X_{mi}(A) + f_{me}X_{me}(A) \tag{58}$$

where $X_{mi}(A)$ characterizes the energetic heterogeneity of the microporous structure and $X_{me}(A)$ describes the energetic heterogeneity of
the mesopore surface.

According to the assumption of the α_s-method [Eq. (3)], θ_{me} is
replaced by the relative standard adsorption θ_r measured for the
reference nonporous solid; consequently, $X_{me}(A)$ in Eq. (58) may be
replaced by $X_r(A) = -d\theta_r(A)/dA$, which describes the energetic heterogeneity of the reference nonporous solid. After this replacement
the following expression for $X_t(A)$ is obtained:

$$X_t(A) = f_{mi}X_{mi}(A) + f_{me}X_r(A) \tag{59}$$

Equation (59) indicates the procedure for calculating the total
adsorption potential distribution $X_t(A)$. The α_s-method provides the
values of a_{mi}^0 and a_{me}^0, which permit calculation of f_{mi} and f_{me}.
The distribution $X_r(A)$ may be calculated from the standard adsorption
curve $\theta_r(A)$; however, $X_{mi}(A)$ may be evaluated from the adsorption
curve $\theta_{mi}(A)$. This procedure has been discussed elsewhere [41]; here
our discussion is focused on the calculation of $X_{mi}(A)$. Figure 12
presents the adsorption potential distribution $X_{mi}(A)$ for the selected
activated carbons from Table 1; these functions show a similar behavior
to those produced by DR Eq. (19) (cf. Fig. 6). Table 6 contains the
values of $\overline{A}$ and σ_A associated with $X_{mi}(A)$ for activated carbons studied.
A characteristic feature of the distribution function $X_{mi}(A)$ is its
widening in the direction of the high values of A, which is especially
manifested for strongly heterogeneous microporous solids [133]. The
important advantage of the values $\overline{A}$ and σ_A associated with the distribution function $X_{mi}(A)$ [Eq. (51)] is the dependence of the ratio $\sigma_A/\overline{A}$
on the parameter n [133]:

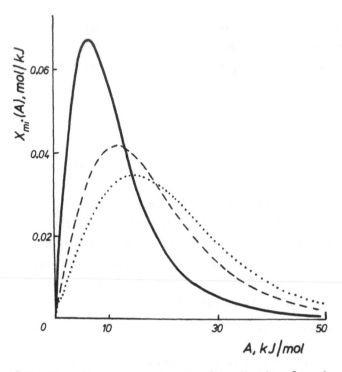

FIG. 12 Adsorption potential distribution function $X_{mi}(A)$ calculated according to Eq. (51) for activated carbons AC (solid line), F-400 (dashed line), and HY (dotted line).

TABLE 6 Values of $\overline{A}$ and σ_A Calculated According to the Respective Eq. (53) and (54) for Activated Carbons from Table 1

Code of carbon	$\overline{A}$ [Eq. (53)] (kJ/mol)	σ_A [Eq. (54)] (kJ/mol)
CWZ-3	19.1	12.1
AC	12.6	11.8
RKD-4	19.7	11.8
F-400	19.1	14.6
BH	12.2	17.7
HY	22.5	14.8

$$\sigma_A/\overline{A} = \left[\frac{4}{\pi}\,\phi_A{}^2 - 1\right]^{\frac{1}{2}} \tag{60}$$

For small values of n, the values ϕ_A differ significantly from unity and the values of σ_A are larger than the suitable values of $\overline{A}$. For values of n greater than about 2, the values of ϕ_A are close to unity and then the ratio $\sigma_A/\overline{A}$ approaches 0.523, which is characteristic for DR Eq. (15). In contrast to DR Eq. (15), Eq. (60) associated with the isotherm Eq. (42) predicts the physically realistic dependence of $\sigma_A/\overline{A}$ on the structural heterogeneity parameter n [133].

C. Total Surface Area of Microporous Solids

The total surface area S_t of microporous solids may be defined as the sum of the specific surface area S_{me} of the mesopores and the geometric surface area S_{mi} of the micropore walls:

$$S_t = S_{me} + S_{mi} \tag{61}$$

Dubinin [113] expressed the geometric surface area S_{mi} of the micropore walls as follows:

$$S_{mi} = a_{mi}{}^0 v_a \int_0^\infty \frac{J(x)}{x}\,dx \tag{62}$$

where v_a is the molar volume of the liquid adsorbate at the adsorption temperature and $a_{mi}{}^0 v_a$ is the micropore volume. Equation (62) for the gamma micropore-size distribution $J(x)$ [Eq. (46)] gives the following expression [136]:

$$S_{mi} = a_{mi}{}^0 v_a \phi_A (q\zeta/n)^{\frac{1}{2}} \tag{63}$$

Table 7 contains the values of S_{mi} calculated according to Eq. (63) for activated carbons listed in Table 1. These values were used to calculate the total surface area S_t according to Eq. (61); the values S_{me} were taken from the α_s-method (cf. Table 2). The calculated values of S_t were compared with those obtained by the BET method (cf. Table 7). This comparison showed that the values S_t are similar to the S_{BET} values; it means that Eq. (61) associated with Eq. (63) provides a simple way for calculating the geometric surface area of activated carbons. Keeping in mind the problems connected with use

TABLE 7 Comparison of the Total Surface Area Calculated According
to Eq. (61) with the BET Specific Surface Area for Activated Carbons
Listed in Table 1

Code of carbon	S_{mi} [Eq. (63)] (m^2/g)	S_t [Eq. (61)] (m^2/g)	S_{BET} (m^2/g)
CWZ-3	970	1136	1340
AC	650	800	1300
RKD-4	810	950	1160
F-400	710	785	1000
BH	270	653	860
HY	460	580	610

of the BET method for evaluating the specific surface area of microporous activated carbons, the method of its evaluation by means of
Eq. (61) seems to be attractive especially for strongly heterogeneous
microporous samples.

VII. THERMODYNAMICS OF GAS ADSORPTION ON MICROPOROUS SOLIDS

A. General Relationships for Thermodynamic Functions

Bering et al. [137,138] derived expressions for thermodynamic functions that characterize adsorption in the micropores. The Gibbs free
energy ΔG^a is expressed by Eq. (11) taken with the minus sign, that
is, $\Delta G^a = -A$. According to Bering et al. [137], the differential
molar entropy ΔS^a is given by

$$\Delta S^a = (\partial A/\partial T)_{\theta_{mi}} + \tilde{\alpha} \, [\partial A/\partial(\ln a_{mi})]_T \tag{64}$$

where the symbol $\tilde{\alpha} = -d(\ln a_{mi}^0)/dT$ denotes the thermal coefficient
of the micropore adsorption capacity taken with the minus sign. The
differential molar enthalpy ΔH^a may be calculated according to the
following equation [137]:

$$\Delta H^a = -A + T \, \Delta S^a \tag{65}$$

where ΔS^a is expressed by Eq. (64).

Equation (64) simplifies significantly if the the curve $\theta_{mi}(A)$ is temperature-invariant, that is,

$$(\partial A/\partial T)_{\theta_{mi}} = 0 \qquad (66)$$

For the temperature-invariant curve $\theta_{mi}(A)$ Eq. (64) gives

$$\Delta S^a = \tilde{\alpha} \; [\partial A/\partial(\ln a_{mi})]_T \qquad (67)$$

Jaroniec [134] showed that Eq. (67) may be expressed in terms of $\theta_{mi}(A)$ and $X_{mi}(A) = -d\theta_{mi}(A)/dA$:

$$\Delta S^a = -\tilde{\alpha}\theta_{mi}(A)/X_{mi}(A) \qquad (68)$$

Substituting Eq. (68) into Eq. (65) we obtain the expression for ΔH^a.

An interesting thermodynamical relationship was derived for the enthalpy of immersion ΔH_{im} of a completely microporous solid [133]:

$$\Delta H_{im} = -\overline{A}(1 + \tilde{\alpha}T) \qquad (69)$$

where A is the average adsorption potential. Substitution of Eq. (20) into Eq. (69) gives an expression associated with DR Eq. (15); this equation was derived by Stoeckli and Kraehenbuehl [139] and initiated studies on the use of calorimetric measurements for characterizing microporous activated carbons [60,65,140]. Expressions for ΔH_{im} involving structural heterogeneity of the microporous solids have been discussed elsewhere [136]. Combination of Eqs. (53) and (69) gives an expression for ΔH_{im} associated with the isotherm Eq. (42).

B.   Thermodynamic Functions for Special Adsorption Models

Bering et al. [137,138] studied the thermodynamic expression associated with the DR Eq. (15). Their expression for ΔS^a may be obtained by substituting Eqs. (15) and (19) into Eq. (68):

$$\Delta S^a = -\tilde{\alpha}\beta^2/(2BA) \qquad (70)$$

However, the LF Eq. (22) and Eq. (23) generate the following expression for S^a:

$$\Delta S^a = -(\tilde{\alpha}/\nu)\{1 + \exp[-\nu(A - \overline{A})]\} \qquad (71)$$

Figure 13 shows a comparison of the curves $\Delta S^a(A)$ calculated according to Eqs. (70) and (71); however, Fig. 14 presents the dependence

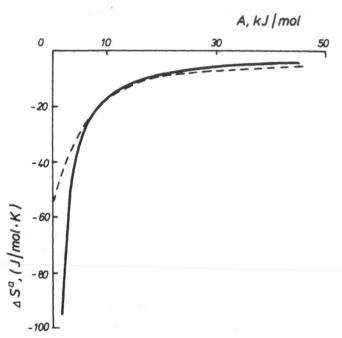

FIG. 13 Comparison of the differential molar entropy curves $\Delta S^a(A)$ predicted by Eq. (70) (solid line) and (71) (dashed line). Calculations were made for $\tilde{\alpha} = 1.022 \times 10^{-3}$ 1/K, $\beta = 1$, $B = 0.003$ (mol/kJ)^2, $\bar{A} = 12.9$ kJ/mol, and $\nu = 0.155$ mol/kJ.

$\Delta S^a(\theta_{mi})$. It follows from Figs. 13 and 14 that the entropy curves behave similarly at moderate and high values of A but differ at low values of A; the curve corresponding to the DR equation increases from minus infinity, whereas the curve relating to the LF equation increases from a finite value. The curves $\Delta S^a(A)$ are increasing functions; however, the $\Delta S^a(\theta_{mi})$ curves are decreasing functions.

Equations for thermodynamic functions of adsorption on heterogeneous microporous solids were discussed by Wojsz and Rozwadowski [141,142] and Jaroniec [134]. Wojsz and Rozwadowski [141,142] discussed the expressions for ΔS^a and ΔH^a associated with Eq. (36). However, Jaroniec [134] derived the expressions for ΔH^a and ΔS^a associated with Eq. (42). For ΔS^a he obtained the following simple equation:

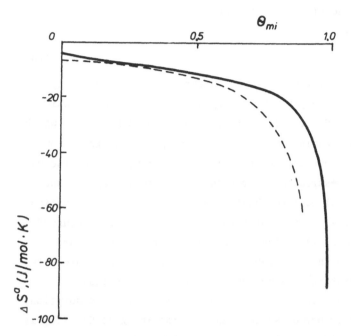

FIG. 14 Comparison of the differential molar entropy curves $\Delta S^a(\theta_{mi})$ associated with DR equation (solid line) and LF equation (dashed line). These curves were obtained from those presented in Fig. 13 by transforming the A axis on the θ_{mi} axis.

$$\Delta S^a = - \frac{\tilde{\alpha}(q\beta^2 + A^2)}{2A(n + 1)} \tag{72}$$

Extensive model studies based on the equations for the gamma microporesize distribution have been presented elsewhere [143]. These studies indicated that the gamma-type distribution is one of the best functions for representing structural heterogeneity of the micropores.

VIII. CONCLUDING REMARKS

The classical DR Eq. (15) is one of the most frequently used equations for characterizing adsorption systems with microporous solids (cf. some recent papers [35,144-148] dealing with this equation). Besides its enormous applications to characterize the microporous structures of activated carbons, it was the object of several modifications

[81,82,149-151].   One of them is the well known Dubinin-Astakhov
(DA) equation [151]:

$$\theta_{mi} = \exp[-B(A/\beta)^{r}] \tag{73}$$

where $r \geq 1$, assumed originally to be integral [2], is frequently
treated as an adjustable parameter independent of the model considerations [152-154].   For $r = 2$ Eq. (73) reduces to DR Eq. (15), which
was found to be applicable for describing adsorption on microporous
activated carbons.   Some authors [116,123,124,155] postulated the use
of DA Eq. (73) for describing adsorption in uniform micropores and
formulated an integral equation analogous to the Stoeckli's equation,
Eq. (28).   Theoretical aspects of this approach have been discussed
elsewhere [156]; this discussion showed that further studies are
desirable to confirm the applicability of DA Eq. (73) for representing the local adsorption on some types of microporous solids.

It was shown that the theoretical description of gas adsorption
on heterogeneous microporous solids based on the integral Eq. (28) is
useful for characterization of microporous activated carbons.   In
particular, application in this description of the gamma-type micropore-
size distribution permitted derivation of the simple equations for the
adsorption isotherm, the adsorption potential distribution, and other
thermodynamic functions that characterize the process of micropore
filling and provide a valuable information about structural and energetic heterogeneities of the micropores.   This description may be
easily extended to adsorption of nondissociating organic compounds
from dilute aqueous solutions on activated carbons [13,89,157-160].
An interesting perspective for the future studies of adsorption on
microporous solids is extension of this approach to adsorption from
nonelectrolytic liquid mixtures.   Further studies in this direction
will permit formulation of a unified theoretical description of
adsorption on heterogeneous microporous solids, which will reflect
similarities and differences between adsorptions from gaseous and
liquid phases.   Up-to-date theoretical achievements in adsorption
from gaseous and liquid phases on energetically heterogeneous surfaces

[13] create foundations for future studies in adsorption on micro-porous solids.

REFERENCES

1. S. J. Gregg and K. S. W. Sing. *Adsorption, Surface Area and Porosity,* 2nd ed. Academic Press, London, 1982.

2. M. M. Dubinin. *Progr. Surface Membrane Sci. 9*:1 (1975).

3. K. S. W. Sing, D. H. Everett, R. A. W. Haul, L. Moscou, R. A. Pierotti, J. Rouquerol, and T. Siemieniewska. *Pure Appl. Chem. 57*:603 (1985).

4. R. T. Yang. *Gas Separation by Adsorption Processes.* Butter-worths, London, 1987.

5. G. E. McGuire. *Anal. Chem. 59*:294 (1987).

6. M. Jaroniec and P. Bräuer. *Surface Sci. Rep. 6*:65 (1986).

7. D. M. Young and A. D. Crowell. *Physical Adsorption of Gases.* Butterworths, London, 1962.

8. D. Dollimore. *Surface Technol. 4*:121 (1976).

9. D. H. Everett, G. D. Parfit, K. S. W. Sing, and R. Wilson. *J. Appl. Chem. Biotechnol. 24*:199 (1974).

10. J. Cortes. *Adv. Colloid Interface Sci. 22*:151 (1985).

11. D. C. Havard and R. Wilson. *J. Colloid Interface Sci. 57*:276 (1976).

12. W. A. Steele. *The Interaction of Gases with Solid Surfaces.* Pergamon Press, Oxford, 1974.

13. M. Jaroniec. *Adv. Colloid Interface Sci. 18*:149 (1983).

14. M. Jaroniec and J. Choma. *Mater. Chem. Phys. 18*:103 (1987).

14a. A. Swiatkowski and J. Choma. *Chem. Stos. 31*:563 (1987).

15. M. Jaroniec and J. Choma. *Mater. Chem. Phys. 15*:521 (1986).

16. M. M. Dubinin. *Carbon 23*:373 (1985).

17. H. Juntgen. *Carbon 15*:273 (1977).

18. H. Jankowska, A. Swiatkowski, and J. Choma. *Activated Carbon.* WNT, Warsaw, 1985 (in Polish).

19. A. Capelle and F. de Vooys, eds. *Activated Carbon . . . a Fascinating Material.* Norit N.V., Amersfoort, 1983.

20. M. M. Dubinin. *Carbon 21*:359 (1983).

21. D. H. Everett and J. C. Powl. *J. Chem. Soc. Faraday I 72*:619 (1976).

22. L. F. Smirnova, V. A. Bakaev, and M. M. Dubinin. *Carbon 25*:599 (1987).

23. K. S. W. Sing. *Berichte Bunsen Gesellschaft phys. Chem. 79*:724 (725)

24. S. J. Gregg and K. S. W. Sing. In *Surface and Colloid Science* (E. Matijevic, ed.), Vol. 9. Plenum Press, New York, 1976, pp. 231-359.

25. P. Bräuer. *Wiss. Z. Karl-Marx-Univ. Leipzig, Math.-Naturwiss. R. 35*:379 (1986).

26. B. C. Lippens and J. H. de Boer. *J. Catal. 4*:319 (1965).

27. K. S. W. Sing. *Chem. Ind. 67*:829 (1967).

28. M. Ternan. *J. Colloid Interface Sci. 45*:270 (1973).

29. T. G. Lamond and C. R. Price. *J. Colloid Interface Sci. 31*:104 (1969).

30. B. Rand and H. Marsh. *J. Colloid Interface Sci. 40*:478 (1972).

31. K. S. W. Sing. In *Surface Area Determination* (D. H. Everett, ed.). Butterworths, London, 1970, p. 25.

32. G. D. Parfitt, K. S. W. Sing, and D. Urwin. *J. Colloid Interface Sci., 53*:187 (1975).

33. R. A. Roberts, K. S. W. Sing, and V. Tripathi. *Langmuir 3*:331 (1987).

34. F. Rodriguez-Reinoso, J. M. Martin-Martinez, C. Prado-Burguete, and B. McEnaney. *J. Phys. Chem. 91*:515 (1987).

35. K. Kaneko, Y. Nakahigashi, and K. Nagata. *Carbon 26*:327 (1988).

36. P. J. M. Carrott, R. A. Roberts, and K. S. W. Sing. *Carbon 25*:59 (1987).

37. M. Jaroniec, R. Madey, J. Choma, B. McEnaney, and T. Mays, *Carbon 27*: (1989), in press.

38. P. J. M. Carrott, J. H. Raistrick, and K. S. W. Sing. *Colloids Surfaces 21*:9 (1986).

39. O. Kadlec. *Collect. Czech. Chem. Commun. 30*:2415 (1971).

40. M. M. Dubinin and O. Kadlec. *Carbon 13*:263 (1975).

41. J. Choma, M. Jaroniec, and J. Piotrowska. *Mater. Chem. Phys. 18*:409 (1987).

42. M. M. Dubinin, O. Kadlec, and N. S. Poliakov. *Izv. Akad. Nauk SSSR, Ser. Khim. 87*:719 (1987).

43. M. M. Dubinin, L. I. Katajeva, and N. S. Poliakov. *Izv. Akad. Nauk SSSR, Ser. Khim. 87*:2410 (1987).

44. M. M. Dubinin and N. S. Poliakov. *Izv. Akad. Nauk SSSR, Ser. Khim. 86*:1932 (1986).

45. M. M. Dubinin and N. S. Poliakov. *Izv. Akad. Nauk SSSR, Ser. Khim. 85*:1943 (1985).

46. M. M. Dubinin and O. Kadlec. *Carbon 25*:321 (1987).

47. S. J. Gregg and J. F. Langford. *Trans. Faraday Soc. 65*:1394 (1969).

48. S. Ali and B. McEnaney. *J. Colloid Interface Sci. 107*:355 (1985).

49. F. Rodriguez-Reinoso, J. M. Martin-Martinez, M. Molina-Sabio, R. Torregrosa, and J. Garrido-Segovia. *J. Colloid Interface Sci. 106*:315 (1985).

50. S. J. Gregg and M. M. Tayyab. *J. Chem. Soc. Faraday Trans. 74*:8 (1978).

51. J. M. Martin-Martinez, F. Rodriguez-Reinoso, M. Molina-Sabio, and B. McEnaney. *Carbon 24*:255 (1986).

52. A. Linares-Solano, J. D. Lopez-Gonzalez, J. M. Martin-Martinez, and F. Rodriguez-Reinoso. *Adsorption Sci. Technol. 1*:123 (1984).

53. H. Jankowska, A. Swiatkowski, J. Oscik, and R. Kusak. *Carbon 21*:117 (1983).

53a. J. Choma, M. Jaroniec and J. Piotrowska. *Carbon 26*:1 (1988).

54. J. Choma, H. Jankowska, J. Piotrowska, and M. Jaroniec. *Monatsh. Chem. 118*:315 (1987).

55. A. Swiatkowski and J. Choma. *Chem. Stos. 31*:563 (1987).

55a. M. Jaroniec and J. Choma. *Materials Chem. Phys. 19*:267 (1988).

56. K. H. Radeke, D. Gelbin, H. Jankowska, and A. Swiatkowski. *Chem. Technol. 36*:247 (1984).

57. M. Jaroniec and J. Choma. *Colloids Surfaces* in press.

58. M. Jaroniec, J. Choma, A. Swiatkowski, and K. H. Radeke. *Chem. Eng. Sci. 43*:3151 (1988).

59. H. Jankowska, A. Swiatkowski and S. Zietek. *Biuletyn WAT (Warsaw) 26*:131 (1977).

60. H. F. Stoeckli and F. Kraehenbuehl. *Carbon 22*:297 (1984).

61. A. V. Kiselev and N. N. Mikos. *Zh. Fiz. Khim. 22*1043 (1948).

62. D. Dollimore and G. R. Heal. *J. Appl. Chem. 14*:109 (1964).

63. J. Jagiello and J. Klinik. *Przem. Chem. 64*:608 (1985).

64. M. M. Dubinin. *Zh. Fiz. Khim. 30*:1652 (1956).

65. K. H. Radeke. *Carbon 22*:473 (1984).

66. M. M. Dubinin, O. Kadlec, N. S. Poliakov, and V. F. Surobikov. *Izv. Akad. Nauk SSSR, Ser. Khim. 87*:1453 (1987).

67. M. M. Dubinin, T. M. Izotova, O. Kadlec, and O. L. Krainova. *Izv. Akad. Nauk SSSR, Ser. Khim. 75*:1232 (1975).

68. M. Jaroniec, A. Patrykiejew, and M. Borowko. *Prog. Surface Membrane Sci.* 14:1 (1981).

69. S. Ross and J. P. Olivier. *On Physical Adsorption.* Wiley-Interscience, New York, 1964.

70. G. F. Cerofolini. *Thin Solid Films* 23:129 (1974).

71. W. A. House. In *Colloid Science: The Specialist Periodical Reports* (D. H. Everett, ed.), Vol. 4. The Chemistry Society, London, 1982, Chap. 1.

72. G. F. Cerofolini. In *Colloid Science: The Specialist Periodical Reports* (D. H. Everett, ed.), Vol. 4. The Chemistry Society, London, 1982, Chap. 2.

73. M. Jaroniec and R. Madey. *Physical Adsorption on Heterogeneous Solids.* Elsevier, Amsterdam, 1988.

74. M. Jaroniec and R. Madey. *Carbon* 25:579 (1987).

75. R. R. Zolandz and A. L. Myers. *Progr. Filtr. Sep. Sci.* 1:1 (1979).

76. W. A. House. *J. Colloid Interface Sci.* 67:166 (1978).

77. J. A. Britten, B. J. Travis, and L. F. Brown. *AIChE Symp. Ser.* 79:1 (1983).

78. L. K. Koopal and K. Vos. *Colloids Surfaces* 14:87 (1985).

79. B. McEnaney, T. J. Mays, and P. D. Causton. *Langmuir* 3:695 (1987).

80. M. Jaroniec. In *Preprints of IUPAC-Symposium on Characterization of Porous Solids.* Dechema, Bad Soden, 1987, p. 62.

81. M. Jaroniec. *Surface Sci.* 50:553 (1975).

82. M. Jaroniec and J. A. Jaroniec. *Carbon* 15:107 (1977).

83. M. M. Dubinin and L. V. Radushkevich. *Dokl. Akad. Nauk SSSR* 55:331 (1947).

84. F. Rodriguez-Reinoso and A. Linares-Solano. *Chemistry and Physics of Carbon 21,* 1-146.

85. D. P. Timofeev. *Zh. Fiz. Khim.* 48:1625 (1974).

86. H. F. Stoeckli. *Chimia* 28:727 (1974).

87. M. M. Dubinin. In *Characterization of Porous Solids* (S. J. Gregg, K. S. W. Sing and H. F. Stoeckli, eds.), Society of Chemical Industry, London, 1979, pp. 1-11.

88. H. F. Stoeckli. *J. Colloid Interface Sci.* 59:184 (1977).

89. M. Jaroniec and J. Piotrowska. *Monatsh. Chem.* 117:7 (1986).

90. M. Jaroniec and R. Madey. *Carbon* 26:107 (1988).

91. D. N. Chakravarti and N. R. Dhar. *Kolloid Z.* 43:377 (1927).

92. M. Jaroniec. *Thin Solid Films* 71:273 (1980).

93. D. Bobok, E. Kossaczky, and M. Sefcikova. *Chem. Zvesti* 29:303 (1975).

94. T. V. Lee, J. C. Huang, D. Rothstein, and R. Madey. *Separation Sci. Technol.* 19:1 (1984).

95. T. V. Lee, R. Madey, and J. C. Huang. *Separation Sci. Technol.* 20:461 (1985).

96. R. Madey, P. J. Photinos, D. Rothstein, R. J. Forsythe, and J. C. Huang. *Langmuir* 2:173 (1986).

97. R. Forsythe, R. Madey, D. Rothstein, and M. Jaroniec. *Carbon* 26:97 (1988).

98. M. Jaroniec, R. Madey, and D. Rothstein. *Chem. Eng. Sci.* 42:2135 (1987).

99. M. Jaroniec. R. Madey, and D. Rothstein. *Pol. J. Chem. 64:* in press (1989).

100. J. R. Sips. *J. Chem. Phys.* 16:420 (1948).

101. B. P. Bering and V. V. Serpinsky. *Izv. Akad. Nauk SSSR Ser. Khim.* 74:2427 (1974).

102. B. P. Bering and V. V. Serpinsky. *Izv. Akad. Nauk SSSR Ser. Khim.* 78:1732 (1978).

103. M. M. Dubinin, T. S. Jakubov, M. Jaroniec, and V. V. Serpinsky. *Pol. J. Chem.* 54:1721 (1980).

104. M. M. Dubinin. *Izv. Akad. Nauk SSSR Ser. Khim.* 78:17 (1978).

105. M. M. Dubinin. *Izv. Akad. Nauk SSSR Ser. Khim.* 78:529 (1978).

106. M. M. Dubinin and T. S. Jakubov. *Izv. Akad. Nauk SSSR Ser. Khim.* 77:2428 (1977).

107. T. S. Jakubov, B. P. Bering, M. M. Dubinin, and V. V. Serpinsky. *Izv. Akad. Nauk SSSR Ser. Khim.* 77:463 (1977).

108. M. Jaroniec and A. W. Marczewski. *J. Colloid Interface Sci.* 101:280 (1984).

109. M. Jaroniec and A. W. Marczewski. *Monatsh. Chem.* 115:997 (1984).

110. T. I. Izotova and M. M. Dubinin. *Zh. Fiz. Khim.* 39:2796 (1965).

111. P. Pendleton and A. C. Zettlemoyer. *J. Colloid Interface Sci.* 98:439 (1984).

112. M. M. Dubinin and H. F. Stoeckli. *J. Colloid Interface Sci.* 75:34 (1980).

113. M. M. Dubinin. *Carbon* 19:321 (1981).

114. M. M. Dubinin. *Carbon* 17:505 (1979).

115. M. Jaroniec and R. Madey. *J. Chem. Soc. Faraday II* 84:1139 (1988).

116. M. Rozwadowski and R. Wojsz. *Carbon 22*:363 (1984).

117. U. Huber, H. F. Stoeckli, and J. P. Houriet. *J. Colloid Interface Sci. 67*:195 (1978).

118. H. F. Stoeckli, A. Perret, and P. Mena. *Carbon 18*:443 (1980).

119. A. Janosi and H. F. Stoeckli. *Carbon 17*:465 (1979).

120. H. F. Stoeckli and J. P. Houriet. *Carbon 14*:253 (1976).

121. H. F. Stoeckli, J. P. Houriet, A. Perret, and U. Huber. In *Characterization of Porous Solids* (S. J. Gregg, K. S. W. Sing, and H. F. Stoeckli, eds.). Society for Chemical Industry, London, 1979, pp. 31-39.

122. H. F. Stoeckli, A. Perret, and U. Huber, *Bull. Chem. Soc. Jpn. 53*:835 (1980).

123. R. Wojsz and M. Rozwadowski. *Carbon 24*:225 (1986).

124. R. Wojsz and M. Rozwadowski. *Carbon 24*:451 (1986).

125. R. Wojsz and M. Rozwadowski. *Monatsh. Chem. 118*:893 (1987).

126. R. Wojsz and M. Rozwadowski. *Monatsh. Chem. 117*:453 (1986).

127. H. Jankowska, M. Jaroniec, and J. Choma. *Przem. Chem. 67*:15 (1988).

128. M. M. Dubinin. *Dokl. Akad. Nauk SSSR 275*:1442 (1984).

129. M. M. Dubinin, S. N. Efriemov, L. I. Katajeva, and E. A. Ustinov. *Izv. Akad. Nauk SSSR Ser. Khim. 85*:255 (1985).

130. M. M. Dubinin, N. S. Poliakov, and E. A. Ustinov. *Izv. Akad. Nauk SSSR, Ser. Khim. 85*:2680 (1985).

131. M. M. Dubinin. O. Kadlec, L. I. Katajeva, G. Okampo, and N. S. Poliakov. *Izv. Akad. Nauk SSSR Ser. Khim. 87*:12 (1987).

132. M. Jaroniec and R. Madey. *Separation Sci. Technol. 22*:2367 (1987).

133. M. Jaroniec, R. Madey, X. Lu, and J. Choma. *Langmuir 4*:911 (1988).

134. M. Jaroniec. *Langmuir 3*:795 (1987).

135. M. M. Dubinin. *Carbon 25*:593 (1987).

136. M. Jaroniec and R. Madey. *J. Phys. Chem. 92*:3986 (1988).

137. B. P. Bering, M. M. Dubinin, and V. V. Serpinsky. *J. Colloid Interface Sci. 21*:378 (1966).

138. B. P. Bering, M. M. Dubinin, and V. V. Serpinsky. *J. Colloid Interface Sci. 38*:185 (1972).

139. H. F. Stoeckli and F. Kraehenbuehl. *Carbon 19*:353 (1981).

140. F. Kraehenbuehl, H. F. Stoeckli, A. Addoun, P. Ehrburger, and J. B. Donnet. *Carbon 24*:483 (1986).

141. R. Wojsz and M. Rozwadowski. *Chem. Eng. Sci. 40*:105 (1985).

142. R. Wojsz and M. Rozwadowski. *Chem. Eng. Sci. 40*:1765 (1985).

143. J. Choma and M. Jaroniec. *Monatsh. Chem. 119*:545 (1988).

144. J. Garrido, A. Linares-Solano, J. M. Martin-Martinez, M. Molina-Sabio, F. Rodriguez-Reinoso, and R. Torregrosa. *Langmuir 3*:76 (1987).

145. J. Garrido, A. Linares-Solano, J. M. Martin-Martinez, M. Molina-Sabio, F. Rodriguez-Reinoso, and R. Torregrosa. *J. Chem. Soc. Faraday Trans. I 83*:1081 (1987).

146. B. McEnaney. *Carbon 25*:69 (1987).

147. S. J. Gregg. *Colloids Surfaces 21*:109 (1986).

148. H. Marsh. *Carbon 25*:49 (1987).

149. S. Ozawa, S. Kusumi, and Y. Ogino. *J. Colloid Interface Sci. 56*:83 (1976).

150. T. V. Lee, J. C. Huang, D. Rothstein, and R. Madey, *Carbon 22*:493 (1984).

151. M. M. Dubinin and V. A. Astakhov. *Izv. Akad. Nauk SSSR Ser. Khim. 71*:5 (1971).

152. B. Rand. *J. Colloid Interface Sci. 56*:337 (1976).

153. K. Kawazoe, T. Kawai, Y. Eguchi, and K. Itoga. *J. Chem. Eng. Jpn. 7*:158 (1974).

154. G. Finger and M. Bulow. *Carbon 17*:87 (1974).

155. M. Suzuki and A. Sakoda. *J. Chem. Eng. Jpn. 15*:279 (1982).

156. M. Jaroniec and R. Madey. *Chem. Scripta 29*: in press (1988).

157. M. Jaroniec, J. Piotrowska, and A. Derylo. *Carbon 18*:439 (1980).

158. M. Jaroniec. *Langmuir 3*:673 (1987).

159. M. Jaroniec. In *Fundamentals of Adsorption* (I. A. Liapis, ed.), American Institute of Chemical Engineering, New York, 1987, pp. 277-285.

160. M. Jaroniec, R. Madey, J. Choma, and J. Piotrowska. *J. Colloid Interface Sci. 125*:561 (1988).

Index

Printed and bound by CPI Group (UK) Ltd, Croydon, CR0 4YY

17/10/2024

01775703-0002